'Tim has pioneered a way and now generously gives us his guidebook, which will save us time and money and bring us to his destination – the future – at a discount.'
Dr Saul Griffith, co-founder Rewiring Australia

'At the national level, there may have been the wasted years of the climate wars; at the household level, it's a different story, and Tim Forcey is shining a (clean, green) light on how we can all act local and think global.'
Zoe Daniel, Independent Member for Goldstein

'I'm convinced that *power to the people* was a phrase devised specifically for this book.'
Damon Gameau, director and producer, *That Sugar Film, 2040*

'Bye-bye winter shivers, summer sweats, dangerous open gas flames and sky-high energy bills. With Tim's expert advice, transform your home into a modern, cosy, budget-friendly oasis. Everyone, from home owners to renters, should be able to live comfortably and affordably, without harming our planet. Tim shows us how.'
Tamara DiMattina, founder The New Joneses and Buy Nothing New Month

'Tim Forcey has written an accessible and entertaining guide on how to electrify your life at home. You don't have to be climate concerned or conscious to benefit from his advice. Forcey shows how you can save money, live more comfortably and play your part in climate action all at the same time.'
Rebecca Huntley, researcher and author, *How to Talk About Climate Change in a Way That Makes a Difference*

T0362873

'A life-changer! In this informative and practical book, Tim Forcey takes readers on an enlightening journey toward transforming their homes and the planet. Houses, apartments, owners, landlords and renters – there's something for everyone. If you want to reduce your environmental footprint and slash your energy bills, this is the ultimate resource you've been waiting for!'
Heidi Lee, CEO Beyond Zero Emissions

'A long overdue resource with tailored options for renters to help make their energy bills lower and their homes more healthy and comfortable.'
Joel Digman, executive director Better Renting

'Tim Forcey's *My Efficient Electric Home Handbook* is a wonderfully well-researched, practical and easy-to-understand guide for those wishing to break up with fossil fuels and transition their homes, businesses and communities to healthier, more comfortable, lower cost and carbon neutral living.'
Brendan Condon, founding director The Cape Efficient
All Electric Community, Cape Paterson, Victoria

'An empowering and informative book.'
Alan Pears, AM, Senior Industry Fellow RMIT University,
Fellow University of Melbourne

'Tim Forcey is the original prophet of efficient electric homes and the benefits they deliver to regular Aussie households. This is the book that brings that good news to the masses.'
Luke Menzel, CEO Energy Efficiency Council

'Essential reading for every Australian.'
Jonathan La Nauze, CEO Environment Victoria

My Efficient Electric Home Handbook

How to slash your energy bills, protect your health & save the planet

TIM FORCEY

murdoch books
Sydney | London

Published in 2024 by Murdoch Books, an imprint of Allen & Unwin

Murdoch Books Australia
Cammeraygal Country
83 Alexander Street,
Crows Nest NSW 2065
Phone: +61 (0)2 8425 0100
murdochbooks.com.au
info@murdochbooks.com.au

Murdoch Books UK
Ormond House, 26–27 Boswell Street,
London WC1N 3JZ
Phone: +44 (0) 20 8785 5995
murdochbooks.co.uk
info@murdochbooks.co.uk

 A catalogue record for this book is available from the National Library of Australia

A catalogue record for this book is available from the National Library of Australia.

ISBN 978 1 76150 033 6

Cover design by Christa Moffitt
Internal and back cover illustrations by Matt Golding
Text design and typesetting by Midland Typesetters
Printed and bound in Australia by Pegasus Media & Logistics

Disclaimer: Before acting on the ideas in this book, please seek professional advice tailored to you and your home.

We acknowledge that we meet and work on the traditional lands of the Cammeraygal people of the Eora Nation and pay our respects to their elders past, present and future.

10 9 8 7 6 5 4 3

The paper in this book is FSC® certified. FSC® promotes environmentally responsible, socially beneficial and economically viable management of the world's forests.

*To my wife of 40 years, the love of my life.
And to my kids and their partners, who continue
to support me through every crazy thing.*

*And to the thousands of people I've met on this journey,
working hard, volunteering hard and donating hard
to make this world a better place.*

Contents

Foreword

From little things big things grow.

The challenges of climate change are a responsibility for us all, not just in how we vote but in what we do. If we are to limit global warming to prevent a future of catastrophe for our children and for the planet, it is up to all of us.

The actions must be global, national and personal.

Scientists have been warning since the 1980s at least that the hydrocarbon economy that emerged with the Industrial Revolution had delivered so much greenhouse gas into the atmosphere that, without change, global temperatures would rise so severely as to make much of the planet uninhabitable and life severely constrained elsewhere.

Since the 1990s one international conference after another has tried to win the quality of pledges from nations around the world to ensure the planet remains liveable, but the lip service offered by most national governments has seen the inexorable rise in global temperatures.

The climate is changing – and for the worse. Sea levels are rising, risking the futures of low-lying island nations, and creating the first generation of climate refugees. There will be more. Storms, hurricanes and cyclones are becoming more frequent and more intense. Bushfires are more intense and are a risk for larger parts of the year. Many of these so called 'natural' disasters I saw before I was an MP, when I was a foreign correspondent covering all manner of untold destruction around the world. And yet, in Australia, a decade of government effectively denied climate change and then delayed action, actively sabotaging the steps that would have reduced our carbon footprint.

The only time, apart from during the COVID years, when Australia stabilised and started to reduce carbon emissions came after the

Gillard government introduced a carbon tax, which experts argue would still be the most efficient and equitable way of expediting the shift to a clean and prosperous economy.

Instead, because of the timidity of one side of politics and the hostility of the other major player, we are left with second-best 'solutions', which may or may not enable us to achieve net zero by 2050.

Our governments are still approving coal and gas projects; our environmental regulations do not have a climate trigger; our laws do not contain a duty of care to future generations.

This makes Tim Forcey's practical work a rare, tangible gift to our communities. It shows us what we can do as individuals to take climate responsibility for ourselves – to show the way by making our homes more energy efficient – and to save ourselves money while contributing to a better world. It enables all of us to do something within our individual means. And it is the efforts of householders that are making a big difference already.

Australia's world-leading take-up of rooftop solar has made a real contribution. As coal-fired power stations close, rooftop solar, along with wind and solar farms, are making up an increasing amount of the power available in the grid. Regular consumers have filled the gap, saved themselves money and made our energy system cleaner and greener.

I am one of the many who stalk Tim's Facebook group, My Efficient Electric Home, for tips. Many of my friends and family are on it too.

Tim has also spent time with me and my husband at our home, helping advise on draught proofing, insulation and the gradual switch away from gas to induction cooking and to heat pumps for heating and cooling. Like most people entering this process, we are electrifying our house as we can afford it and taking steps to make it more energy efficient along the way.

At the national level there may have been the wasted years of the climate wars; at the household level it's a different story, and Tim Forcey is shining a (clean, green) light on how we can all act local and think global.

Zoe Daniel, Independent Member for Goldstein

Introduction: why live in an efficient electric home?

Would you like your home to be more comfortable in winter and summer, and cheaper to operate throughout the year, all without costing the earth?

It's easy to answer that question with a quick 'yes', but the next one is harder. Do you have a good idea how to go about it?

And there are so many questions after that! What's the most affordable and cleanest way to heat a home? Is insulation important or should you focus on your windows? What's a heat pump? If you draught-proof your home, how will you let in fresh air?

Based on my experience working with thousands of households, this handbook will help you convert your home into a more affordable all-electric place of comfort.

In this book I describe how to take advantage of technologies, including how to heat your home with a reverse-cycle air conditioner, use a heat pump to heat your water, cook with an electric induction cooktop and get the most out of solar photovoltaic (PV) panels.

Put a different way, this handbook shows how you can stop burning stuff at home, whether that be fossil fuels such as gas or LPG, or even wood.

The 'efficient' part of this book's title relates to how you'll find tips to ensure your home is suitably draught-proofed and insulated, well ventilated, has appropriate window coverings, and more.

But is it really worth going to the trouble to do all of this? Why would you want to invest the money and personal effort to change your home when, even though it might not be perfect, it's not that bad? Here are some reasons:

* You can save a lot of money. At our home, we're saving $3,000 every year.
* Your home can be more comfortable – not too hot and not too cold.
* Your home can be healthier and safer for you and your family.
* We're in a climate emergency and must get our homes off fossil fuels as quickly as we can.

That last reason has motivated me for nearly twenty years. I'd like there to be a safe climate future for my children and grandchildren, as well as for your children and grandchildren.

Getting our homes off fossil fuels is one thing we can do to slow down climate disruption. Whereas staying on fossil fuels allows the gas industry to say, 'You must really love the stuff, let's go frack for some more!'

And here's some good news. Most Australian governments at the local, state and federal levels now agree with the idea of electrifying our homes. They are providing rebates, credits and other incentives to help us get our homes off fossil fuels. This assistance can make converting cheaper than you might think.

That's a win-win-win-win!

Crying out for help

I call myself a home comfort and energy adviser. My wife thinks that sounds weird, but it's the best description I could come up with. Over the past fifteen years I've chatted with thousands of Australians in their homes, at their dining tables and kitchen benches, helping them work out a home improvement plan.

Why do people eventually pick up the phone and ask me for help? Usually, it's because they're too cold. Or they're too hot. Plus, these days, more people are working from home and are reminded every day how uncomfortable their homes really are.

'My house is draughty, the windows are freezing and the insulation, if we have any at all, must be poor. I'm tired of suffering!' they say.

My clients also tell me their gas and electricity bills have exploded! And they wonder about the climate emergency and getting off fossil fuels. 'I'd like you to help me get my home off gas.'

There is also the worry about using gas around young children, and other home health concerns. 'My child has asthma.' Or 'We've seen a bit of mould around this winter. I've heard I shouldn't hang my laundry in the lounge room to dry, but what else am I supposed to do?'

And then there's the new tech. 'Should I get solar panels on my roof? Or a battery? Or an electric car? And what's a heat pump?' These are common questions.

So, there are lots of reasons why people ring me. They're uncomfortable, confused, not sure what to do next, what their top priority should be or how they will get the best bang for their buck.

But not everyone can ring me, and this information needs to be spread as widely as possible. That's why I've written this book.

What you'll find in this handbook

This book is divided into two parts.

The first part, 'Why this book?' describes how I grew up on a dairy farm, then worked for 30 years as a chemical engineer in the fossil fuel industry. I explain how my exposure to the climate impacts of fossil fuels led my wife and me to start a climate action group and volunteer with others. Eventually I changed careers, did some research at the University of Melbourne, started a popular Facebook group and, ultimately, began my own home comfort and energy advice business.

I describe the steps I took to remove fossil fuels from our own 120-year-old weatherboard home in Melbourne, converting it to a comfortable, energy saving, better than net-zero greenhouse gas emission, efficient electric home.

I then cover the bigger picture of why we should move all our homes away from gas and other fuel burning and explain how our electricity supplies are becoming greener every day.

In the second part, 'How to create your own efficient electric home', I've included my detailed tips about how you can use electrical appliances to:

- actively heat and cool your living spaces (with heat pumps, aka reverse-cycle air conditioners)
- heat your water (also with heat pumps)
- switch from cooking with gas to cooking with electricity
- and then how to totally disconnect from the gas grid.

This is followed by tips on how to improve your home via:
- draught-proofing
- air quality and moisture management, including clothes drying
- insulation
- window coverings, inside and out
- improved window frames and glazing.

After that, to complete the all-electric picture, I explain how to:
- monitor and minimise your home's electricity use
- generate your own electricity with solar PV panels on your roof
- decide if it is yet time to invest in home batteries.

Is this book only for home owners? Not at all. Much can be applied in rentals and I summarise this in the last chapter.

Help from a few thousand friends

In this book I share my learnings from working with my clients in their homes. Further, I summarise the countless discussions I've had with householders online at the 116,000+ member public Facebook group, 'My Efficient Electric Home (MEEH)' which I started in 2015.[1] At MEEH, we've influenced many millions of dollars' worth of home-improvement decisions.[2] We add to that total every day. We've even had an influence on which homes some members decided to buy in the first place.

The MEEH Facebook group is more than just a chat, it's a searchable online database. It's grown to be the world's largest store of information on improving home comfort and energy performance. But it's a fast-flowing river that many find tricky to navigate. That's why, here, I've tried to tame that river and provide the key information you need to get started.

I believe this book is the first and perhaps only of its kind written by an experienced home comfort and energy adviser who has been in thousands of homes and has helped thousands more people online. The messages in this book have been road-tested many times over the last decade, making for a comprehensive and accessible handbook.

Let's get started

Every action we take to reduce our dependence on fossil fuels will help our Earth by reducing greenhouse gas emissions. In the process, we can also make ourselves healthier and more comfortable, and our homes more affordable. So, if you'd like to:

* electrify your home/get it off gas
* make your home healthier and more comfortable for you and your family
* shrink your energy costs
* make your gas bill disappear
* get the most out of renewable energy and storage technologies, such as solar panels, air conditioners, heat pumps, batteries and electric vehicles

... this is your handbook!

I hope you enjoy reading it.

(Please note, where I mention commercial products or suppliers, I have no commercial connections with any of them; my advice is independent. They are simply products and suppliers that seem to do the job.)

Part 1
Why this book?

1.

The climate emergency calls for change

In this chapter I describe why I stopped working in the fossil fuel industry and instead began to work out ways for us to thrive without that industry. I also examine why our housing is generally unfit for purpose, harming our health as well as our wallets.

From dairy farmer to fossil engineer

Where I can, I like to keep things simple. Go with what works. But I'll also do what has to be done. You can try ignoring things that matter for a little while, but eventually they'll bite you on the bum.

Who taught me this approach? It might have been my father. He was a dairy farmer on a small, unprofitable plot in the sometimes-freezing, coal-filled mountains of Pennsylvania in the US. Or it might have been my mum, who was a nurse. I don't suppose nurses 'mess around' any more than struggling dairy farmers do.

When faced with one of those first global energy crises around 1973, my parents took action. I recall being down in the musty basement wrapping our electric hot water service with insulation to keep the heat in the tank. We even replaced some of our old draughty windows with triple-glazed glass, way back in the 1970s.

Another energy crisis hit in 1979. I learned to drive while focusing on the fuel efficiency display on our pick-up truck. I learned about the fuel-economy impact of fully inflating the tyres. Even then the world was trying to educate me about energy.

This background led me to study chemical engineering at university. How could I personally fix the energy crisis? And as an additional benefit, given the global needs of industry to respond to energy price shocks, a well-paid career path beckoned that promised to show me the world.

Thus, I ended up in Melbourne, Australia, working first in petrochemicals and later in fossil oil and gas. With mining companies ExxonMobil and BHP, I'd fly offshore to visit remote oil and gas platforms in the Gulf of Mexico, the Persian Gulf, the Timor Sea and offshore Victoria out in the Bass Strait. Stimulating times!

Among other things, I developed a deep understanding of the role gas plays in Australian homes, businesses, industries and in electricity generation.

Along came the climate emergency

But there was a problem in the fossil oil and gas industry, one we started to hear about in 1992, the year of the first United Nations climate conference. We learned that fossil fuels had a use-by date.

As a chemical engineer, I eventually figured out what we're doing with our Earth is one massive uncontrolled chemistry and physics experiment. When I was working in the chemical industry, we controlled runaway chemical reactions to ensure we didn't ruin the product, blow up equipment, kill plant workers or poison the community – all very bad outcomes. But that's exactly what we are doing with our Earth's atmosphere. We're also cutting down and burning trees, which are important carbon stores and vital for removing greenhouse gas from our atmosphere.

Unbelievably, as late as 2023, our political leaders are still allowing miners to search for and develop more fossil fuels. This ensures damaging greenhouse gas emissions will spew from these new sources for decades.

I remained working with BHP's petroleum division until 2010. Up until I left, management would reflect that at least the oil and gas impacts might not be 'as bad as coal'. Eventually I could no longer stomach the unwavering enthusiasm they had for getting fossil fuels out of the ground and burning them as fast as possible. Today, nearly a decade and a half later, I wonder about the psychological state of the people who keep working in fossil fuels. I guess they won't buy this book.

In my last fossil-fuel years I did try to effect change, as they say, 'from within'. I was accepted into former US Vice President Al Gore's Climate Reality Project, where I learned more about the climate crisis and solutions. I presented my findings to upper BHP management and other business leaders, to politicians, and to school and community groups, as I still do today. The demand for such information is only increasing. People are keen to know what they can do.

Learning from volunteers

Back in those days, not-for-profit organisations with which I volunteered, such as Beyond Zero Emissions (BZE) and Renew (formerly the Alternative Technology Association), sought to inform the Australian public and politicians about climate solutions. More recently, they have been joined by the notable Dr Saul Griffith and his organisation Rewiring Australia.

I'm also aware of the hundreds of active community groups across Australia, from the dynamic Electrify Boroondara, to Electrify 2515 (a postcode in the Illawarra region of NSW), to my local Melbourne Bayside Climate Crisis Action Group (BCCAG). This group, which my wife and I helped found in 2006, was where I first started offering my neighbours free home advice after receiving volunteer training organised by Bayside Council. These volunteer-led community groups continue to help Australian households move beyond fossil fuels. They also encourage politicians to respect the climate science and listen to voices beyond the cashed-up fossil-fuel industry.

After BHP, I worked for two years at the Australian Energy Market Operator (AEMO). As you will read more about later, AEMO is the official quasi-government organisation responsible for managing Australia's electricity grids and parts of our gas distribution networks. I started there as the 'gas guy' because that was my background.

I finished at AEMO in 2012 after leading their first 100 per cent renewable electricity planning study. I was recognised within AEMO as knowing more about renewable energy than most of the engineers there at the time, thanks to my previous volunteer work with BZE. At BZE we were two years ahead of AEMO when we published Australia's first detailed 100 per cent renewable plan for decarbonising Australia's electricity grids: the 'Zero Carbon Australia Stationary Energy Plan'.[1]

Discovering the pros of going gas-free

In 2013, the team at BZE (where I continued as a volunteer) went on to publish another groundbreaking report and book: the 'BZE Buildings Plan' and *The Energy Freedom Home.*[2] We showed how all of Australia's buildings – our homes included – could operate without using any fossil fuels.

A couple of years later, I had a role as a casual researcher at the University of Melbourne Energy Institute. There we produced research reports looking at a range of topics, including pumped hydro energy storage, trends in electricity demand, and the methane emissions associated with coal seam gas extraction and fracking. And we did one more thing which, looking back, has led to me writing this book. In 2015, we were the first to show how Australian households could save hundreds of millions of dollars every year.[3] How? By heating their homes and water using heat pumps, which in Australia we often refer to as 'reverse-cycle air conditioners'.

Heat pumps are cheap to operate because they use a refrigerant system to extract free renewable heat from the thin air outside your home. A nice thing about free heat is that it is free. No one can charge you for it. The arrival of this cheap building and water heating technology caught many by surprise in 2015 and still does today.

> A nice thing about free heat
> is that it is free.

By 2015, the price of gas across eastern Australia had gone up dramatically while, at the same time, the capabilities of reverse-cycle air conditioners just kept getting better and better. To me this was the biggest consumer win-win-win ever: a way to reduce the cost of living, while improving household health and safety and getting our homes off fossil gas and reducing greenhouse gas emissions.

When we released our research, we received some media attention.[4] But we needed other ways to broadcast the message of how we have cheaper and cleaner ways to heat our homes.

Scaling up in-home climate action

One way to spread this information was via social media. I created a public Facebook group and named it 'My Efficient Electric Home (MEEH)'. (Since the name of that group is very similar to the title of this book, I'll use the acronym MEEH when referring to the Facebook group.)

Initially, among the 50 or so friends and other original members, we discussed ways to get the most out of our air conditioners. But the topics soon spread to cover all aspects of home comfort and energy: insulation, draught-proofing, solar panels, hot-water heat pumps, moisture management, etc., right down to oven seals and doggy doors.

The membership at MEEH grew and grew, and it continues to grow today. We passed 100,000 members in August 2023. That's 50,000 more members than we had the year before. Two hundred or more people may join on a single day. I learned a lot staring at MEEH online nearly every day for its first few years. I became intensely interested in the things people were doing in their homes to reduce pollution, reduce energy bills and to become more comfortable.

Becoming an adviser

Professionally, I also continued to learn a lot when I moved from the University of Melbourne to the not-for-profit Moreland Energy Foundation (which later became the Australian Energy Foundation or AEF). There, I assisted Victorian government and local council-funded programs by visiting people's homes to offer advice or to identify potential upgrades. We visited rich households and poor, and everything in between. We visited renters, home owners and people in publicly funded housing.

At the AEF I was involved with the development of the Victorian government's computer-based home rating tool, known as the Residential Efficiency Scorecard, and became a qualified 'Scorecard' assessor.[5]

SEE ALSO *Do you live in a leaky bucket? (page 122) for more about the Scorecard*

Once I combined

* what I learned when visiting people in their homes
* what I learned from thousands of online discussions
* the principles I knew from chemical engineering, such as refrigeration, radiation, air flow and heat transfer
* what I had learned in my other jobs about gas, electricity and renewable energy technologies, markets and distribution networks

I began to independently advise people about how they could improve their homes.

At first this consisted of offering free advice online, but then people started to ring me up with more and more questions. Finally, my wife said, 'You know, people would pay you for this.' So, I started my own business in, of all years, the bushfire- and COVID-plagued 2020. Since then, I've enjoyed visiting more than 600 homes so far in Melbourne, regional Victoria and Canberra.

What's a home comfort and energy consult?

Across Australia, there are hundreds of people trained to offer home comfort and energy advice. Each adviser or assessor will, to some extent, have their own techniques and methods and offer varying services.

When I work with clients in their homes, I describe it as being like a tutorial. With the home owner or renter, I look at the comfort and energy aspects in the home and all related structures and equipment, such as heating and cooling systems, windows, opportunities for zoning living spaces, bathroom extraction fans, air conditioner filters, clothes drying options and much more. I make suggestion after suggestion about how things could be improved.

We look in the roof space for insulation, we crawl under the floor, we measure the water flow from the shower heads, we examine the energy bills, we see if the solar panels are dirty or if they are, in fact, functioning at all, we check the age of the hot water service and so on.

A session like this generally goes on for three hours. Hopefully my clients learn a lot about their homes. I also learn something new in just about every home I visit.

Before I leave a client's home, we go back through the suggestion list. We work out the top priority items for the short and longer term. We discuss which items will have the best bang-for-buck: the most benefit for the least effort or least money spent. What are the exact next steps? Who are the reliable suppliers and installers the client can ring next to get a quote for, say, insulation, a hot-water heat pump or draught-proofing services? What fixes can the client attempt DIY (do-it-yourself)?

 If you want to find a Scorecard assessor for your home, go to homescorecard.gov.au/find-a-scorecard-assessor or timforcey.com.au.

People find it ironic that although I migrated to Australia to work in the fossil oil and gas industry, these days I advise folks on how they can get off the stuff. And why not, because today we have far cheaper and cleaner options.

At the University of Melbourne, I published articles covering home energy economics at The Conversation and Climate Spectator. Since then, I've continued to publish articles for Renew Economy, The Fifth Estate, *Renew Magazine*, Crikey and elsewhere. I have presented to community groups, contributed to a few podcasts, conducted numerous radio interviews and have briefly appeared on television. Type my name into your search engine along with any key term or phrase such as 'heat pump', 'gas disconnection', 'double-glazing' or 'methane' and you'll find my work.

Why are Australian homes so poor?

Australian homes are notoriously poor performers when it comes to keeping energy bills down. And what can be even sadder is that they also perform badly in terms of keeping us healthy and comfortable.

Stories of Canadians or Swedes saying things like, 'I was never so cold until I came to Melbourne', are common. Medical data highlights that many premature deaths, particularly of the elderly or infirm, can be linked to poor home performance during the depths of winter or the hottest nights of summer.[6]

But why is this becoming so noticeable only now? Perhaps in bygone years we were less concerned with indoor and outdoor air quality given our preoccupation with cigarette smoking and our cars' reliance on leaded fuel. Maybe we made do with adding an extra jumper if we were cold or slept on the verandah if we were too hot. We didn't plan to live until we were 90 because reaching our 70s was considered a 'good innings'. We didn't consider the impact fossil fuels had on our climate.

So maybe we asked less of our homes back then. But also, Australian homes have a way of not telling you they have a problem.

I have lived in both icy, snowy conditions and in humid, subtropical conditions. These were climates no one would call 'mild'. We'd run either the heating or cooling 360 days out of 365. And one thing that happens in climates like that is the home itself will tell you if it wasn't built properly. For example, an uninsulated wall can lead to ice forming inside a bedroom. Snow might blow in from beneath an unsealed front door and leave a charming little snow drift piling up in your hallway. If water seeps into a structure, the subsequent ice freezing and then thawing can tear a home apart in one season. In subtropical conditions, mould and wood rot can get out of hand very quickly. Build a poor-performing home in these sorts of climates and it will soon become obvious, well within the warranty period, that something needs to be fixed.

Whereas in the mild climate zones that cover much of Australia, it may be many years before a home's lack of insulation is identified or the cause of a draught coming in from under a kitchen cabinet is revealed. Our homes might not scream out in any obvious way, 'Hey, I've got a problem.' Although at times, we can leave it to the local fauna (and in extreme cases even the local flora) – a mouse plague, a carpet snake or a possum in the roof – to tell us where our builder left some holes.

In our relatively mild Australian conditions, it's easy for builders to take shortcuts, as the problems may not appear for a long time.

What about new-build homes?

My focus is on existing homes because working with clients in their homes is where I gained most of my experience. Whereas with the homes my wife and I have owned, although I've worked with designers and builders as we undertook renovations, I've never built a new home from the ground up. Nevertheless, many of the points I make in this book can be applied to new builds.

While I do review plans for clients building new homes, this is only a small part of my professional work. I have to say it's a part that I often find unrewarding. It's disappointing to me (but no longer surprising) to see how little attention is paid to comfort and energy matters by the professionals my clients have hired (e.g. architects, designers and builders). Sadly, time and budget constraints mean that many of my recommendations, though of keen interest to the client, won't be set right for years.

So, the first word of advice is this. If you're thinking of building a new home or embarking on a major renovation, choose your professionals carefully.

But more positively, I have also met many architects, designers and builders who *are* switched on to home comfort and energy matters.

*If you're thinking of building a home,
choose your professionals carefully.*

Home energy rating tools can be used to assess both new-build designs and existing homes. With these you can employ building science to see how your new home will perform long before construction has even started.

SEE ALSO *Do you live in a leaky bucket? (page 118) for more on how to use these tools*

Should you leave the electricity grid?

The things I describe in this book will apply to most Australian homes – homes that are destined to remain connected to our electricity grids.

I recommend that as soon as you can, you electrify your home and vehicles. Ditch the fossil gas grid and the petrol grid. Ditch the diesel grid, the LPG grid and the wood grid!

However, your home and lifestyle *are* likely to remain connected to the electricity grid, as well as to the water and sewerage grid, and to a bunch of other things that we might not think of as 'grids', such as the internet and telecommunications grid, the food and clothing grid and the job grid.

It would be a very expensive exercise to shift an all-electric home, accompanied by one or two electric vehicles, off the electricity grid altogether. You would end up in an inconvenient and compromised place if you tried.

Our electricity grids will continue to provide large benefits by supplying electricity to our homes when we're unable to make enough ourselves, and also allow us to share the excess electricity we generate with our neighbours. That's the world we're building.

Rejecting ideas from overseas and the past

I've lived overseas, as have many Australians. In other countries we might see something that makes sense in homes there and wonder, 'Why can't we do this in Australia?' However, please be careful, because not everything that makes sense overseas will make sense here.

Some examples that I explore in this book, but that you'll find I don't support, include:

* hydronic heating
* thermal mass
* Passivhaus construction methods.

Likewise, our older Australian homes contain a record of things that were done in the past. Have good ideas been forgotten? Or is it time to move on and embrace new ways?

Some examples of past practices that I explore and reject include:

* burning fuels in and around the home, which includes cooking with gas
* uncontrolled air leakage
* evaporative cooling (in increasingly humid climates)
* hanging laundry around the home to dry on the rainiest winter days
* solar-thermal water heating systems
* cooling roof spaces with ventilating equipment
* skylights.

2.
Electrifying our homes is now a thing

If you take some of the actions described in this book, your home will become more comfortable, healthier and cheaper to operate. Your home will predominantly be 'fuelled' by renewable heat and renewable electricity, and you will have reduced your home's impact on our environment and climate.

In this chapter, I discuss how electrifying our homes and getting them off gas is growing in popularity. Then I discuss the steps, using my home as a case study.

A decade in the making

At the time of writing this book, we are now a decade along from when the volunteers and staff at Beyond Zero Emissions (BZE) showed how our homes could be electrified. Back then, when we looked around for examples of electrified homes, there weren't many to be found. The homes of BZE colleagues Matthew Wright, Richard Keech and John Shiel were leading examples.[1]

Ten years later, it's easy to say that electrifying and 'degassing' homes has become a 'thing'. Reaching more than 116,000 members in the MEEH Facebook group is one indication. The development of hundreds of local community groups that are helping households to electrify across Australia is another, as is the expansion of local council, and state and federal government activity in this area. Businesses are springing up with names like 'Goodbye Gas', 'Pure Electric' and 'Electrify Victoria'.

Entire new housing estates are being built with no connection to any gas grid,[2] and sometimes with restrictions on burning LPG or wood. Dozens of local councils, led by the more progressive ones such as Merri-bek in the northern suburbs of Melbourne,[3] have proactive home-electrification programs. The government of the Australian Capital Territory (ACT) is enacting policies to get homes off gas.[4] The Victorian government has declared that as of 1 January 2024, new homes that require a planning permit will not be connected to the gas grid.[5] Independent federal parliamentarians are pushing for all Australian states to follow suit.[6] The Australian government announced in its May 2023 budget that $1.3 billion will be spent supporting home energy-efficiency retrofits and electrification.[7]

The gas distribution businesses are being asked by their former gas customers to remove gas meter after gas meter from their residential properties. These distributors have submitted funding requests to the regulator so they can start decommissioning the gas distribution networks.[8]

What's driving change?

The climate emergency is one factor driving people to change. For example, we saw thousands of new members join the MEEH Facebook group during the 2019–2020 bushfire and consequent smoke crises.

The quick increase in gas prices is another reason. In winter, I've visited Melbourne homes where people are spending $15, $20 and even $25 a day to heat their homes by burning gas. An afternoon presenter on Melbourne 3AW radio said he spends even more: $30 a day over winter.[9] Australian gas is no longer a cheap fuel.

Solar panels on Australian roofs have been popular for fifteen years now and are a visible way we see our neighbours taking action. But once the solar panels are up, people next want to get the most out of the electricity they generate. This leads to householders learning about better ways to heat and cool their homes, heat water and cook their meals.

Finally, perhaps people are just tired of being uncomfortable in their homes. Too cold in winter and too hot in summer. They've heard it doesn't have to be this way, so many are ready to invest in home improvements.

Admittedly, not everyone who wants to improve their home is able to make immediate progress. In addition to lacking information (which this book aims to help with), a poll I ran identified eleven barriers to people electrifying their homes, with access to money/financing being a key barrier, unsurprisingly.[10]

Our home's fifteen-year transition

I'll use the conversion of our old weatherboard home as a case study. However, I'm well aware that our home wasn't the first to electrify. Far from it. Many had acted before we did. Eventually we got there.

And now there are thousands of other households around Australia who have done the same and are enjoying the benefits. Tomorrow there will be even more homes added to the all-electric list. Some fortunate new home owners are moving into efficient electric homes from the very beginning, and they won't need to do any conversion. Lucky them.

In 1994, we moved into our old Edwardian weatherboard home in Melbourne's Bayside. The house had been built in 1903. Immediately we needed to add a second floor consisting of two bedrooms and a bath for our three young children.

There were decisions to be made about the windows with our extension. Migrating as I did from an overseas land of ice and snow, I asked, 'Do we need double-glazed windows?'

'Nah,' the builders said, 'you don't need double-glazed windows.' Okay. Back then, I'm not sure I could have found such windows in Melbourne anyway.

'How about insulation in the walls?' I asked. The answer was the same. 'Nah, you don't need insulation in the walls.' On that one, I pushed back and said, 'Tell you what, how about we put some insulation in the walls.'

And then to complete the extension, we relocated the 'old banger' through-the-wall air conditioner from one room to another, ripped out an old gas-burning wall heater and installed the latest thing, ducted gas heating. Ooh-la-la, it was all the rage back then.

Working in the gas industry, I knew that gas was cheap. And though by then we had heard about climate change, it wasn't front of mind.

Our home's electrification in one chart

We proceeded to do not much with our house for the next fourteen years while we watched the kids grow up. Then we started a long and drawn-out comfort and energy makeover in 2008, which is represented in the chart below.[11]

The black columns in the chart show how in 2008 we used 13,000 kilowatt-hours of gas (equivalent to 47,000 megajoules for those who prefer gas industry units). Along with that, shown by the dashed columns, we bought 6,000 kilowatt-hours of electricity.

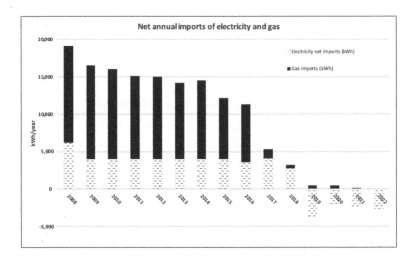

On the chart, the first change you can see happening is in 2009 when we installed one of those early six-panel, 1-kilowatt solar photovoltaic (PV) systems. These were heavily subsidised by federal government rebates. Any electricity that we produced but did not use was fed back into the electricity grid. For this we received a credit from our electricity retailer (this is known as a feed-in tariff). With an attractive feed-in tariff of $0.60/kilowatt-hour in those days, we recouped our heavily government subsidised $499 outlay in about one year.

We obtained those panels as part of a community bulk-buy led by local entrepreneur Erik Zimmerman.[12] I recall walking down the street to ask my neighbours if they were interested in joining in. Many did, although some resisted. One fellow asked, 'What's a small system like that going to do, power my hand drill?' Having never gone door-to-door trying to 'sell' anything other than Girl Scout cookies, that was a lesson for me. You can show people ideas that will benefit their home and our communities and that will pay off financially in as little as one year, but you can't force people to take these ideas on board.

Though a 1-kilowatt solar PV system might be only one-tenth the size of what you might install today, you can see from the chart that it had a measurable impact in reducing the amount of electricity we had to buy from the grid.

Improving our home's 'thermal envelope'

When thinking about heat gain or loss, I refer to the roof, walls, floors and windows of a home as the 'thermal envelope'.

Through 2009 to 2015 we modernised the front and back of our house and, where we could, improved our home's thermal envelope. These renovations included double-glazed windows (removing some of the single-glazed windows we installed in 1994), good insulation, draught-proofing and some interior window treatments.

One thing I failed to add, because it seemed too hard at the time, was underfloor insulation. Our weatherboard home is supported by stumps, but it lies low to the soil beneath. So, if I wasn't writing a book right now, I might be digging under my house, still trying to put in underfloor insulation.

With those improvements, you can see on the chart that the black columns – the amount of gas we were buying each year – get shorter as the years pass. Insulation and those other upgrades work! Our home became more comfortable while at the same time we had to buy less and less gas for winter heating.

Out with gas-fired heating, in with our reverse-cycle air conditioners

With 2017 came the game changer. This was when I finally took my advice from my own University of Melbourne research that it's far cheaper to heat a home with reverse-cycle air conditioners instead of using gas. Why is it so cheap? Using a small amount of electricity to run their fans and refrigerant compressor, these air cons are 'heat pumps' that extract free heat from the thin air outside your house.

We installed two 'split-system' air cons at either end of our house and said goodbye to the ducted gas heating. Thanks to this upgrade, our gas purchases in 2017 were a small fraction of what they had been the year before.

But then there was the other side of the equation: how much did our electricity purchases increase because we were running those air conditioners day after day, all winter long?

The answer was: not much.

As the gas bill disappears, the electricity bill goes up only a little

The gas industry says we shouldn't electrify our homes because it would severely increase household electricity demand. However, this isn't true. It's not what we've found in our house, nor is it what thousands of recently electrified households across Australia are observing, nor is it what we are seeing on our electricity grids. The demand for grid-supplied electricity has actually been falling, while more and more homes have electrified, and even more electric vehicles are seen on our roads every day.[13] How can this be?

There are a number of reasons, including:

* new electrical appliances are much more efficient than the appliances they replace, which reduces electricity demand
* modern reverse-cycle air conditioners are very efficient
* the blower component of a ducted gas heater uses a lot of electricity
* as some households have gotten off gas, they've also installed their own solar PV panels, reducing their need for grid-supplied electricity
* there is an uptake in insulating, draught-proofing and improved window coverings, which reduces the need for home heating and cooling
* as our Earth heats up, less heating is needed each winter.

It's rather amazing, sort of like a gift, that a reverse-cycle air conditioner, a device we might buy to help us survive a sweltering summer, is also by far our cheapest way to heat. However, it's not magic; it's heat pump technology.

SEE ALSO *Heating and cooling (page 61)*

Heating our water with a heat pump

From 2017 to 2018, we reduced our gas burning further when we replaced our 23-year-old gas-fired water heater with a hot-water heat pump. What's that? I explain this in Chapter 6 (see page 93) but, in short, a hot-water heat pump is like a reverse-cycle air conditioner in heating mode. It extracts free heat from the air but in this case puts this heat into water. With a hot-water heat pump, you can heat your water for one-quarter the cost of using a simple electric-resistive hot water service (like an electric kettle), or at one-third the cost of a gas-fired hot water service.

Further, for those homes lucky enough to have solar PV panels on the roof, a hot-water heat pump timed to run during sunlight hours can use the electricity you generate on your own roof to provide perhaps the cheapest hot water in the world.

More solar PV: How much can we fit on our roof?

Toward the end of 2018, it was time for the next big energy move at our home. We added an additional 22 solar PV panels to our roof, increasing the panel capacity from 1 kilowatt to 7.7 kilowatts. The small 1-kilowatt system we installed in 2008 continues to produce today, into its sixteenth year.

Today, 7.7 kilowatts of solar PV capacity is not large. An average-sized system these days is up around 10 kilowatts. But for us, a capacity of 7.7 kilowatts covers the logical unshaded east-, north- and west-facing surfaces of our roof.

On the chart on page 21, then, you can see that our house became, over the course of 2019 and during the following years, a net-annual electricity producer. In other words, at any given moment during a day throughout the year, our house may be importing electricity *from* the grid or exporting electricity *to* the grid, but on a net basis over a full calendar year, our home is a net electricity producer. Basically, over one year at our house, we sell more energy than we buy. Or to say it another way, on a net annual basis, our house isn't a drain on Australia's energy system, it's an asset.

In a moment I'll discuss what that means in dollar terms and how this has reduced our energy bills.

Note that for the years shown in the chart, we have no home battery, nor do we have any electric cars. Electric bicycles, electric lawn mowers, yes, we have those small things. We also have a luxury item, a large electrically heated spa, which uses quite a lot of electricity when it's being heated. If we ditched the spa, our numbers would look even better.

Our kitchen goes gas-free

Part way through 2021, despite COVID-19 lockdowns in Melbourne, we managed to arrange a minor kitchen renovation. We replaced our gas cooktop with a permanent, in-built electric induction cooktop.

Prior to the kitchen renovation, we occasionally used a portable induction cooktop. Now, we use that portable hot plate to cook things outdoors from time to time. We also loan the portable to friends and neighbours who wish to learn more about induction cooking.

On the chart on page 21, you can see we used a small amount of gas for cooking in 2021, but for 2022 the house was gas-free and disconnected from the gas grid.

Finally, we had an all-electric home. Again, many households reached this destination long before we did. The purpose of describing our case study and of writing this book is to help many more households to quickly follow.

Saving $3,000 a year

In 2022, our electricity bills totalled about $600 for the full year. Since our house is connected to the electricity grid and will remain so, we pay about $1 a day for that connection. So that's a $365 yearly bill that we can't avoid.

However, we no longer get a gas bill. So happily, that's $0 for gas. Nor do we burn wood or anything else in the home. During the period shown in the chart, we did not have a home battery or electric cars. Therefore, we still bought petrol for our vehicles, but that's not included in the chart.

So that's $600 a year for our home's total energy costs (petrol excluded).

But what if we had not made any changes back in 2008? What if we had never installed solar PV panels, or what if we were still using gas for heating, hot water and cooking? What if we had never draught-proofed, nor upgraded some windows, nor added some insulation and better-performing window blinds?

At today's energy prices, if our home now was the same as in 2008, we'd be receiving gas and electricity bills totalling $3,600 each year. Our investments in home improvements, which come with health and comfort benefits too, are saving our household $3,000 a year.

These large savings are being replicated around Australia. Every home and the people who occupy it makes for a different situation, but there is no doubt that significant dollar savings are possible when you electrify your home. Beyond money, there are the health and comfort improvements too. I will discuss these benefits in the following chapters.

3.
Stop burning stuff!

Before I describe actions you can take to fully electrify your home, in this chapter I explain why it is an excellent idea to stop burning combustible fuels. These fuels may be in the form of:

* reticulated fossil methane gas (i.e. 'natural' gas)
* reticulated hydrogen or biomethane gas (aka 'renewable' gases), should these become available in the future
* wood or coal
* bottled liquefied petroleum gas (LPG or propane)
* other liquid fuels, such as kerosene or ethanol.

In short, burning fuels such as these:

* is generally far more expensive than using electrically powered options, such as heat pumps (e.g. reverse-cycle air conditioners), and can contribute to additional greenhouse gas loading of our atmosphere, leading to accelerated climate breakdown
* can negatively impact the regional communities from where they are extracted (e.g. the communities in Queensland and New South Wales where coal seam gas is being extracted in harmful ways)
* can pollute the environment inside your home, leading to bad health outcomes for you, and especially impacting small children

* pollutes the neighbourhood outside your home, which can lead to bad health outcomes for you and your neighbours
* can be dangerous if fires and open flames are not controlled
* requires ongoing checking and maintenance to maintain safe performance, adding further to householder costs.

I've got a saying in the MEEH Facebook group: 'It's where we don't burn stuff!'

COOKING WITH GAS (MASKS)

Fossil methane gas is expensive, polluting, unhealthy and dangerous

As I mentioned, I worked for years in the gas industry, overseas as well as in Australia. I am aware that we continue to use a lot of gas in our Australian homes. Historically gas, especially in eastern Australia, was seen as a cheap fuel for heating homes and many other things. In fact, up until about 2015, we enjoyed the cheapest gas in the developed world. Why was this gas so cheap?

Eastern Australian gas used to be cheap

When oil and gas companies go exploring, they'd prefer to find oil rather than gas. Being in liquid form, oil is more valuable than gas, more useful than gas, not to mention easier to ship around the world. Companies such as BHP/Esso in the Bass Strait, Santos at Moomba in north-east South Australia and Woodside in Western Australia, indeed found oil in the late 1960s and into the 1980s, but along with the oil came gas.

Australia quickly became awash with 'stranded' gas. The question was: what to do with it all?

In some parts of the world, gas found with oil may be thrown away, either by being vented raw into our atmosphere with massive climate consequences or by being burned off (i.e. 'flared'), which also has large climate consequences.

For the Bass Strait and Moomba, rather than allowing the by-product gas to be tossed away, Australian federal and state governments required that a market be found for it. Thus, pipelines were laid from the oil and gas fields to what would eventually become markets in Melbourne, Adelaide and Sydney. Later, pipelines connected all the central and eastern states and territories together, extending 3,000 kilometres from Tasmania, New South Wales and to Queensland, and then over to the Northern Territory.

Western Australia remained separate. Gas there was mainly liquefied, pumped into boats and exported overseas. A small fraction of the gas produced was sent south via a 1700-kilometre pipeline to Perth and Bunbury.

As the gas pipelines came to the capital cities and smaller distribution lines spread into the regions, we found lots of ways to use this gas, including burning it inefficiently in our homes.

Gas remained cheap in eastern Australia for nearly 50 years, for a number of reasons. The oil and gas companies were profiting from the oil sales alone. They didn't need to make any money from gas. They just wanted to get rid of it. And initially, governments capped the price that the gas producers could charge, meaning prices couldn't be increased. And lastly, eastern Australia was a captive buyers' market. Unlike in Western Australia and the Northern Territory, until 2015 there was no way to export eastern Australian gas overseas to where buyers would have been happy to pay more.

But this was about to change.

Scraping the bottom of the barrel

Gas produced from 'legacy' gas fields, such as Bass Strait, Moomba or the North West Shelf in Western Australia, is known as 'conventional' gas. Explorers drill a well looking for oil and if they are successful, they also usually find some gas. That's the conventional way.

Those legacy fields contained wells numbering 'only' into the hundreds. This may seem like a lot of wells, but it was nothing compared with what was to come as the oil and gas industry became more desperate to maintain production.

There are only so many Bass Straits, Moombas and North West Shelfs. That's the reason these Australian oil and gas discoveries date back to the 1960s, '70s and '80s and that there haven't been similar large discoveries since. But a lack of success using conventional techniques did not cause oil and gas producers to stop in their tracks, pack their bags and go home. Instead, they turned to increasingly complex and invasive methods to essentially scrape the bottom of the oil and gas barrel.[1]

Hydraulic fracturing or 'fracking' is one such invasive technology. This is where water, chemicals and sand are injected into the ground at high pressure, opening and widening any natural cracks from which gas and oil can be extracted. Fracking has demonstrably catastrophic consequences, including polluting waterways, releasing dangerous hydrocarbons and increasing ground-level ozone levels, which can harm people with asthma or other respiratory illnesses. The harmful consequences are such that fracking has often been featured in the international news.[2] This has led to justifiable concerns in regional communities. For fracking to stand any chance of being profitable, it requires that thousands, even tens of thousands, of wells be spaced across the countryside, turning lands used for farming and grazing, as well as forests and bushlands, basically into a pincushion.

There are more than one million fracking wells in the United States alone. In Australia, Victoria and Tasmania have outlawed fracking, but the Northern Territory, Western Australia, South Australia, New South Wales and Queensland still allow the practice.

Another invasive technique developed over the past twenty years is coal seam gas extraction, which has had an even bigger impact than fracking on the eastern Australian gas situation.

It's long been known that methane gas collects in coal seams. This has been a major hazard for coal miners ever since coal was first dug out of underground mines. But up until the past twenty or so years, no one had managed to work out a way to produce large volumes of gas from coal seams buried deep underground and, given that there is no oil contained in these coal seams, most conventional oil- and gas-producing companies weren't that interested.

Eventually the clever coal seam gas mining specialists developed technology where tens of thousands of wells were drilled across the countryside, water was pumped out of coal seams and then largely thrown away. With the water gone, eventually gas flowed into the wells.

New Queensland gas-export industry drives up gas prices

Given the vast amount of coal that lies under parts of Queensland and New South Wales, once the gas miners got the hang of it, they realised they could tap into greater volumes of gas than had ever been produced from the Bass Strait or Moomba. It was clear such gas volumes could never be absorbed by the small domestic East Coast gas market. As far as the gas industry was concerned, along with the government bodies who agreed to the plan, this gas had to be exported to the large overseas markets.

Therefore, billions of dollars were spent building gas processing plants in Gladstone, Queensland. There, the gas is liquefied, pumped into ships and sold overseas. Starting in 2015, the new access to international buyers turned the East Coast gas market on its head.[3] What for decades had been an isolated buyers' market offering the cheapest gas in the developed world, rapidly became an export-oriented sellers' market where the wholesale price of gas doubled, doubled again and then went up some more.

> ### Cheap gas is never coming back.

This rapid escalation of gas prices heavily impacted large industrial gas users and shocked the households that had burned a lot of gas for years. It even drove up the price of electricity, given that a small but influential fraction of Australia's electricity is made by burning gas (as I describe in the next chapter, page 48).

Further shocks were to come in 2022 with the conflict in Ukraine further driving up international and domestic gas prices.

The result: gas in eastern Australia is no longer cheap. Cheap gas is never coming back.

Do you know where your gas will come from, and at what price?

If your household continues to use gas, be aware that it is being mined in increasingly damaging ways. It will be coming from farther and farther away and will continue to escalate in price.

The legacy eastern state gas fields of the Bass Strait and Moomba are in their last stages of depletion.[4] More and more of the gas used in the eastern states will be coal seam gas coming down the pipelines from Queensland.

Gas importers are even developing plans to import expensive liquefied gas by ship into ports such as Adelaide, Geelong and Wollongong.[5] This is an idea that just a few years ago was viewed as being crazy.[6] This shows how much the gas market has turned. It's not known if gas imported into Geelong or Wollongong would come from Queensland, Western Australia, the Middle East or even from Russia. What we do know is that imported gas won't be cheap.

As gas must come from farther and farther away, it's becoming more challenging for the gas companies to ensure reliable mid-winter supply to the millions of homes still on gas. AEMO, the quasi-government planning organisation involved with ensuring the gas pipelines running beneath Melbourne streets remain full, posts warnings each winter saying that gas supply may not be guaranteed. There may come periods when large gas users have to reduce their consumption in order to ensure gas flows throughout the suburbs.

Gas grid death spiral

And then there is the so-called 'death spiral' of the gas distribution grid, which works like so. As more and more homes leave the gas distribution networks, this leaves fewer and fewer households to pay for the care and upkeep of the pipelines under all our streets. A dwindling customer base means the gas distributors

are charging each remaining household more and more for the service. The ever-escalating cost per remaining household provides a greater incentive for the next household to leave the gas grid, which once again increases costs for the remaining few. And so on.

Eventually there might be only two groups of households left on the gas grid. There will be those on one end of the income spectrum who have such financial means that acting on energy prices hasn't yet felt necessary. And on the other end, households with limited financial means or other constraints that delay them electrifying their homes.

Governments and regulators across Australia are now thinking about what this means. One thing we know is that the cost-per-household of distributing gas is going only one way – up.

Methane is a powerful greenhouse gas mined in more damaging ways

The gas piped into our homes is mostly made up of the chemical methane. Unfortunately, unburned methane is often released or leaks into our Earth's atmosphere during the stages of production, transport by pipeline and end use.[7] Intentional releases and leaks are disastrous because methane is a greenhouse gas that tonne-for-tonne can be 100 times more damaging than that other well-known greenhouse gas, carbon dioxide (CO_2).

When methane is burned, carbon dioxide is the resulting chemical product (along with water). So, either way, whether methane is released or burned, it has a significant climate impact.

In recent years, a number of methane-spotting satellites have been launched into orbit. These satellites are identifying devastatingly large methane releases from irresponsible oil and gas production operations around the world. Australia is no exception, with studies finding that the methane emissions from fossil gas production are significantly under-reported by the gas industry.[8] Personally, I have travelled to the Queensland coal seam gas fields and witnessed constant venting of methane gas.[9]

Additional methane also gets released, either by accident or design, as the gas travels through pipelines and compressors to reach your home.

Many of us have smelled and reported gas leaks in our streets. Ironically, my wife and I recently had to be evacuated from our all-electric home when a construction crew ruptured the gas main in front of a neighbour's property. Fortunately, there was no fire or explosion. The gas distribution companies across Australia are constantly upgrading their leaking pipelines and other equipment.

Once gas arrives in our homes, gas appliances release more unburnt methane. Gas cooktops release methane directly into our living spaces (along with other chemical contaminants). So-called 'instantaneous' (aka tankless, or continuous-flow) gas-fired water heaters release a large puff of unburned methane each time they produce hot water.

And then, finally, when we buy gas, the first thing we do is set it on fire. Combustion converts most of the methane to carbon dioxide, another greenhouse gas. However, combustion is never 100 per cent complete, meaning even more methane goes into our atmosphere unburned.

The gas system leaks from one end to the other. Given the ongoing climate emergency, we need to stop buying gas, burning gas and producing methane and carbon dioxide emissions as soon as possible.

If gas wasn't in our homes already, it wouldn't be allowed

Burning gas in and around our homes can be unsafe and is definitely not a healthy thing to do. How can gas harm us? Let me count the ways.

1. All gas equipment must be treated with care to minimise the risks of gas fires and explosions. Gas space heaters must be checked no less than every two years by a qualified gas appliance technician to ensure deadly poisonous carbon monoxide (CO) gas isn't entering the air you breathe in your home. Paying for the visit of the gas technician is a cost you'll avoid when you get your home off gas.

SEE ALSO *Heating and cooling (page 61)*

2. If you use a gas cooktop, it's releasing a range of contaminants into your home. To minimise the risk of asthma and/or other difficulties, open windows in the kitchen when cooking (if you can), ensure kitchen extraction fans (if you have one) are in good working order and use these diligently along with an open window to improve fan performance. Ensure that your extraction fan takes air and

contaminants from your kitchen and blows them outdoors, rather than merely recycling them around your rangehood (if this is possible).

SEE ALSO *Bye-bye, gas grid: cooking without gas (page 103)*

3. You might think that burning gas in appliances installed *outdoors,* such as via a ducted gas heater or hot water service, might cause less trouble. However, burning gas always produces a range of polluting chemicals, such as carbon monoxide, sulphur dioxide and nitrous oxides. These chemicals can cause localised health impacts, especially if combustion gases from your own equipment or your neighbour's make their way into your home through an open window. Gas combustion contaminants contribute to the city-wide formation of smog.

Have I got a deal for you!

Let's imagine you're sitting in your fully electrified home. One day you hear a knock on the door. It's a salesperson keen to tell you about a new home energy source. It's a hazardous pressurised gaseous chemical, a flammable and global-warming gas. It must be handled carefully lest it cause burns and explosions. It could set fire to your home or to your entire neighbourhood. You can burn it to heat your home, but it might produce an odourless but deadly chemical that will kill you in your sleep. You can use it to cook your food, but it could also burn you, contribute to a family member contracting cancer, or cause your child to suffer from asthma.

> Given that we have cheaper all-electric options for cooking and heating, if gas wasn't already prevalent under our streets, there's no way we'd allow it to come anywhere near our homes and families.

Renewable gases: a ploy to delay home electrification?

Producers of fossil fuels around the world are under stress. Their products are the main cause of climate breakdown. What's a fossil-fuel business to do?

In Australia, gas producers and distributors are realising that the pace of home electrification is increasing.[10] Their revenue streams are under threat. How will they stay in business? Will they disappear, as did the producers of asbestos and film cameras?

One idea that the gas industry lobbyists have come up with is to suggest we should all keep burning fossil gas in our homes while we wait for them to deliver something that might not be as damaging. That something is called 'renewable gas'.

What's a renewable gas? It could either be biomethane or green hydrogen. Unfortunately, it's become clear that the idea of renewable gases being widely distributed to homes is nothing more than a diversionary tactic aimed at slowing home electrification. Using fossil gas today is more expensive than home electrification. Renewable gases will be even more expensive than fossil gas. And just like fossil gas, renewable gases also have damaging health, safety and climate/environmental impacts.

This idea of renewable gas isn't new. Renewable gases have been talked about for the past twenty years and gas companies will probably still be talking about them twenty years from now.

If it made economic sense to use biomethane or hydrogen in our homes today, we'd already be doing it. But we're not. The high cost of renewable gases places them well beyond the realm of being a realistic fuel for homes.

What is biomethane?

Biomethane can be made from materials such as sewage, food wastes, animal wastes, crop wastes or forest materials.

In the same way that fossil gas mainly consists of the chemical methane, biomethane also mainly consists of the chemical methane. Therefore, biomethane is compatible with and readily interchangeable with fossil gas in pipelines and appliances. (As we'll see in the next section, such interchangeability isn't the case when it comes to hydrogen.)

The concept of biomethane being renewable or climate/carbon neutral is due to its source, which is generally a crop or food waste material originally grown by extracting carbon dioxide from the air. It could therefore be considered carbon neutral if the production process were a complete cycle. For example:

* crop material is converted to biomethane
* biomethane gas is piped into homes
* biomethane is burned to heat and cook, which converts the biomethane back to carbon dioxide, which then drifts into our atmosphere to help another plant to grow, thus completing the cycle.

All this can be possible – if none of the biomethane produced is ever released or leaked during its manufacture, transport, storage or use. Releases and leaks of biomethane must be minimised because it still contains the chemical methane, which is a powerful greenhouse gas, as I discussed earlier.

Unfortunately, zero release or leakage of biomethane is unrealistic and, in fact, impossible. Therefore, biomethane is not a zero-carbon climate solution.

Nor will biomethane be any safer or healthier to have running under our streets or to use in our homes than the fossil gas it is meant to replace. This is because biomethane has nearly the same chemical composition as fossil gas, with all the same fire, explosion and health hazards.

It is also not a cheap fuel. The process for making biomethane is well known and already practised in Australia at a small scale, as well as at a larger scale overseas in countries such as Germany. The costs of production are known and they are high, meaning that biomethane can't compete with the costs of home electrification.

Nor will there be any great quantity of biomethane available for use in our homes. Compared with the amount of fossil gas that we have traditionally used, there isn't that much crop waste or sewage available to make a lot of biomethane. In our future low-carbon world, any limited amount of biomethane that can be made will be used for higher-value purposes, such as making chemicals. We won't be wasting biomethane by burning it in homes.

We have far cheaper, healthier and safer electric options that are, in fact, already being rapidly deployed in our homes. Who is the gas industry trying to kid? Us and governments, yes. But themselves too, I think.

What's all the hype about hydrogen?

Hydrogen is another flammable gas chemical that can be used as a fuel. The fossil gas industry suggests that someday hydrogen could be piped into homes either as pure hydrogen or in a mixture with methane (fossil gas) or biomethane. However, it's now been shown around the world that this was simply a tactic by that industry to delay home electrification and to preserve business-as-usual.[11]

Green renewable hydrogen can be made from renewable electricity and water. This process is well known and has been used for decades, although ongoing research and deployment seeks to improve it and reduce the production costs.

There are many barriers to green hydrogen being used in homes. Most importantly, green hydrogen is far more expensive than fossil gas. If fossil gas can't compete economically with home electrification, hydrogen is even less competitive.

SMITH & SON*
GREEN HYDROGEN PLUMBING
*WHO BELIEVES THIS BUSINESS IS JUST A DIVERSIONARY TACTIC TO SLOW HOME ELECTRIFICATION

GOLDING

Rather than using renewable electricity to create green hydrogen, it is far more useful to use renewable electricity to power a heat pump to heat our homes or water. In fact, it would take *six times* as much renewable electricity to heat a home via the hydrogen route than is required via the heat pump route.

Transporting hydrogen to our homes from its place of manufacture would require huge expense as the gas pipeline network would have to be upgraded, along with metering and control equipment.

But wait, next comes the really hard part: the simultaneous same-day change-out of all of the gas-burning appliances in all of the gas-burning homes and businesses in a given suburb. Imagine an army of thousands of gas plumbers marching in to change out all the gas appliances in a single day. I'd like to see that!

There are even more problems with using hydrogen in homes.

When hydrogen is burned in air, the primary combustion product is water. Unfortunately, asthma-promoting contaminants are also produced, including nitrous oxides. In other words, cooking with a hydrogen cooktop will have health consequences of a similar nature to cooking with methane gas.

Nor is hydrogen a perfect climate solution. When released or leaked into the atmosphere, hydrogen can impact our climate as a secondary greenhouse gas, contributing to climate breakdown.

Green hydrogen will have an important role to play in the industrial manufacture of chemicals and in mineral refining. But it will be too expensive to use in our homes when compared with the common electrical options we are already deploying.

I hope I've made it clear why, at present, nowhere in the world uses hydrogen in homes in any substantial way, and why there's nowhere in the world where there are firm plans to do so. Sadly, hydrogen has been co-opted by the gas industry as a way of deflecting rightful enthusiasm away from home electrification. However, most recently it appears that more and more decision-makers around the world have seen through the fossil gas industry's hydrogen hype.

Wood burning: a growing health issue

As the price of gas goes up, some households are turning to burning wood to keep warm in winter, even right in the centre of Melbourne, a city of more than five million people. In rural areas, wood burning is a long-standing practice, both for home heating and the primordial ambience a fire can impart.

Unfortunately, wood is not a clean burning fuel, and this is a growing health concern. Medical studies have shown there is no level at which it is safe to breathe wood smoke.[12]

I suppose in the good old days when cigarette smoking was common in our homes, offices, planes, and trains, adding a wood fire into the mix wasn't a big concern. However, these days we have higher health expectations. We're not so content to die at 60 years of age from a heart attack, as did my smoking and coal-mining grandfather, nor even at 70 from lung cancer and emphysema, as did my smoking father and mother.

Fortunately, instead of spending our effort to find where we can rustle up some timber for the wood heater and also where we can dispose of the ashes, reverse-cycle air conditioners provide a far more convenient and cost-effective way to heat our homes. In Canberra, for example, where the wintertime wood-burning air pollution problem is well known, residents are being paid to exchange their wood heater for an efficient air conditioner.

And what about burning coal in the home? It's been a while since I've been in an Australian home where coal briquettes were lit up, but coal is probably still being burned somewhere in our suburbs. Like burning wood, burning coal or charcoal-based solid fuels will contaminate both your home's indoor air as well as your neighbourhood. These days we have better options.

LPG and other liquid fuels: too expensive and still polluting

Liquefied petroleum gas (LPG) is produced from oil and gas wells, along with oil and gas. It's another fossil fuel that creates greenhouse gases when you burn it, so I'll call it fossil LPG.

For home use in Australia, LPG largely consists of the chemical propane. It's often used where homes aren't connected to a reticulated methane gas grid. It's stored at the home in large, pressurised metal bottles for which you pay a monthly rental fee. When the bottle is empty, the supplier exchanges it for a full one. LPG is, of course, widely used for outdoor barbeques. LPG can also be used in converted cars and utes as a cheaper automotive fuel than petrol or diesel.

For significant home use, LPG is an expensive fuel when compared with electric options. Heating your entire home with LPG would be an expensive thing to do – almost like burning petrol to heat your home. Using a reverse-cycle air conditioner for living-space heating will be far cheaper than burning LPG.

Likewise, using a hot-water heat pump would be a far cheaper option than burning LPG to heat your water.

And lastly, for cooking, burning LPG unfortunately results in similar health and safety concerns as burning methane gas, including increasing the risk of children developing asthma in their homes.

As the traditional sources of Australian fossil LPG are depleted (e.g. the Bass Strait and Moomba oil and gas fields), Australian households that keep using it might not realise that their LPG is coming from farther and farther away, such as from the Middle

East or Russia. However, they might notice that the price of LPG is climbing even higher. Getting a home off LPG and on to electrical appliances can be an economic no-brainer.

Recently, the fossil LPG supply industry has started talking about a greener LPG option that they call 'bio LPG' or 'renewable LPG'.[13] My concern is that the fossil fuel suppliers are using the idea of a greener LPG as another tactic to delay home electrification, keep people burning fossil fuels and preserve their polluting business-as-usual. It's not clear when, if at all, bio LPG will appear in Australia, nor how much will be available.[14] Its greenhouse impact is also yet to be determined. What is known, however, is that the processing required to create this synthetic product will mean it will be an expensive fuel perhaps suited only for the rich and famous. And again, it will come with the same health and safety problems as we see with methane gas and with fossil LPG. If at present you use LPG in your home, my advice is to work out how to fully electrify your home. Don't be caught out waiting around for bio LPG.

Other liquid fuels, such as kerosene and ethanol, aren't widely used in Australian homes, and it's best we don't start using them. They are expensive fuels that, when burned, can cause health and safety issues inside the home, just like the other combustible fuels I've already described.

Cost-of-living crisis or opportunity?

The combined situation of highly priced gas, imported to us over increasing distances, along with damaging health impacts, and less certainty around continuity of supply in colder months, could be seen as a crisis.

Or, it could be seen as an opportunity to rapidly electrify our homes, ditch an expensive, polluting and dangerous fuel source and to adopt healthier renewable alternatives.

4.
Where our electricity comes from, now and in the future

No longer burning fossil fuels in and around our homes can be the cheapest, healthiest and safest way to set up our homes for the future. But what happens to greenhouse gas emissions and climate impacts when we electrify? And can we be sure that the price of electricity, like gas, won't continue to spiral upwards? In this chapter I explain where our electricity comes from now and in the future.

Our electricity gets greener every day

Australian electricity has long been created by burning fossil coal and gas, and much still is. Does it make sense to electrify our homes before our electricity supplies are completely switched over to be renewable and green?

The answer to this question is *yes*, for the following reasons:

✳ Our grid-supplied electricity is becoming greener (more wind- and solar-based) every day. In 2023, 39 per cent of Australian electricity came from renewables, such as wind, solar and hydro.[1] In South Australia, from time to time, renewables already cover more than 100 per cent of electricity demand (with the excess exported at those times to other states).

✱ By using solar PV panels on the roof of your home, at least during sunlight hours, you can power your electrical devices with renewable electricity that you've generated yourself.

✱ Electrical appliances such as heat pumps capture renewable heat and are far more energy-efficient than gas-burning appliances. Heat pumps need not use that much electricity.

✱ Home energy-efficiency improvements (e.g. draught-proofing, insulation and window coverings) go hand-in-hand with electrification and moderate any electricity demand growth. The study of our home that I presented in Chapter 2 (see page 20) is a case in point. As we got our home off gas, we invested in improvements that turned our home into a net-annual electricity exporter.

The last two points I cover in Chapters 9 to 11. Right now, I'll cover the first two points in more detail.

As I described in the previous chapter, any gas we continue to use in our homes will be increasingly costly, will come from farther and farther away, and have increasing climate, societal and environmental impacts, whereas the electricity we generate across Australia is getting greener every day, derived more and more from the renewable energy sources such as the sun and the wind.

Merely twenty years ago, the use of a small amount of renewable hydroelectricity (water power) meant that only around 4 per cent of Australian electricity came from renewable sources. The other 96 per cent was made by burning fossil coal and gas.

Turning to 2023, the use of renewables had increased to 39 per cent, made up of 19 per cent from solar generation, 13 per cent from wind generation and 7 per cent from hydroelectricity. This was up from 35 per cent the year before.[2] As you electrify your home, you are investing in equipment such as air conditioners and hot-water heat pumps that may last twenty years or more. With ongoing investments in renewables and energy storage, imagine how green our electricity will be in twenty years' time. In fact, the current Australian government has indicated that our electricity will grow from 36 per cent to 82 per cent renewables by 2030.[3] At the time of writing, that's only six years away.

The Australian Energy Market Operator (AEMO) modelled one scenario where electricity generation reaches 96 per cent renewable by 2042.[4] AEMO points out that for Australia to achieve this, we will continue to need excellent government policy, as well as rapid and large investments by governments, businesses and individuals in more solar and wind generation, and in energy storage and electricity transmission lines.

Already South Australia achieves greater than 100 per cent renewable electricity generation from time to time, allowing that state to export excess renewable electricity to Victoria. Tasmania, with its relatively large hydro-electricity capacity, has for many years been able to exceed 100 per cent renewable electricity generation, also exporting its excess to Victoria.

All of these figures take into account the renewable electricity contributed by homes with solar PV panels on their roofs.

Increasing renewable electricity production

Looking beyond annual results, there are hours each year in many locations across our electricity grids where wind- or solar-generated electricity could have met 100 per cent of electricity demand, but inflexible coal-fired power stations continued to operate. Individual homes with solar PV systems that are 'export limited' will be well aware of so-called 'grid constraints', but this happens at the largest scale too. At certain times (for example, a sunny and windy spring Sunday when electricity demand is low), large wind farms are forced to sit idle, while elsewhere on the grid coal continues to burn (albeit at minimum possible levels).

AEMO's plan is that with further investment in transmission lines, energy storage and more solar and wind generation, every Australian coal-fired generator will eventually stop running. As that progressively happens coal plant by coal plant, we'll see constraints on renewable electricity generation lessen. Until then, any extra electricity demand – additional electricity we use in our homes as we burn less gas, as well as when we charge our electric vehicles – will, at times, be met by renewable electricity that otherwise would have never been produced.

Electricity prices: going up or down?

While electricity prices are constantly in the news, gas prices aren't so much. This surprises me, given the huge increase in gas prices we've seen in the last decade across most of Australia. But for some reason, in comparison to electricity prices, the price of gas seems to go under the radar. I'm not sure why this happens.

The headline of a media article might shout 'Power bills going up'. The use of the word 'power' refers to electricity. But then it turns out the article actually covers rising electricity and gas bills. Rather than power bills, a more inclusive term for the headline would be energy bills.

So why is the rising price of gas often ignored? Is it because not every Australian home burns gas? Is it because the biggest gas bill, which usually covers the winter period, comes only once a year? Is it because gas is often billed over a shorter time period than electricity and so a two-month gas bill may seem smaller than a three-month electricity bill?

Whatever the reason, it's ironic that what's caused electricity prices to go up over the last decade has, in fact, been the rising price of gas. Although Australian electricity has in the past mainly been generated by burning coal and more recently by renewables, a smaller amount of electricity is generated by burning gas. Nevertheless, in our electricity markets the use of gas often sets the wholesale market price. The end result is that rising gas prices over the last decade have driven up wholesale electricity prices.

In 2022, due to the conflict in Ukraine, there was a large (but temporary) additional increase in the price of Australian wholesale electricity. That overseas conflict lifted the price of black coal and gas globally. Then in 2023, after a one-year lag, Australian retail electricity prices followed suit with a large increase.

This lag in retail price adjustments is unfortunate because although wholesale Australian electricity prices have come back down in 2023, that price fall is yet to be passed on to households that buy electricity on the retail market.

In future, I expect to see the ongoing roll-out of renewable energy and energy storage decouple Australian wholesale electricity prices from the erratic international fossil fuel prices. Renewables have the advantage that once a wind or solar generator is built, the ongoing running costs are very low, especially when compared with the constant need to buy fossil fuels.

The situation in the Australian Capital Territory (ACT) is a good case study. Years ago, visionary policy-makers declared the ACT would achieve 100 per cent renewable electricity supplies. The ACT achieved this goal in 2020, and therefore folks living there avoided the 2022 fossil fuel price impacts caused by the Ukraine conflict. Lucky to be living in the ACT.

Households taking action

Whether electricity prices go up, down or stay the same, there are many things we can do in and around our homes to reduce our electricity use, make our own electricity or have electricity supplied by our retailer at the best possible price. I cover these and other tips in Chapters 12 and 13.

It is a remarkable fact that in Australia today we require less electricity from our electricity grids than we have at any time since 2004.[5] One reason for this is that newer appliances use less electricity than older versions.[6]

Speaking the language!

While I'm on the topic of electricity, it's important to understand the terms used to measure your gas usage, your electricity usage or production, the electrical power usage or production of your appliances or solar panels, and the differences between all these.

You buy gas by the megajoule (MJ)

Let's start with the gas-industry unit of energy measurement known as the megajoule (MJ).

When you buy gas, you'll pay around two to four cents for every megajoule. Your gas meter is actually measuring the volume (cubic

metres) of gas that goes through the meter, but on your gas bill, the metered volumes are converted to the megajoules of gas energy used. This conversion calculation includes a factor that reflects the chemical composition of the gas coming down the pipeline buried beneath your street, which can change from month to month.

The measured volume of gas is then multiplied by the unit cost of gas energy (quoted in cents per megajoule) to arrive at the amount you pay for the gas you use.

On your gas bill, you'll also see the fixed or daily supply charge in dollars a day. This is a charge for having your home connected to the gas grid, even if you use only a very small amount of gas, or even no gas at all.

If the gas industry wanted to, they could choose to sell gas measured not by the megajoule but rather by the kilowatt-hour. The kilowatt-hour is just a different way in which to measure an amount of energy. The mathematical conversion factor is this: it takes 3.6 megajoules (MJ) of energy to equal 1 kilowatt-hour (kWh) of energy.

You buy (and sell) electrical energy by the kWh

A kilowatt-hour is the electricity-industry unit of energy measurement.

When you buy electricity, you'll pay twenty, 30 or more cents for every kilowatt-hour of electrical energy your home uses. If you have solar PV panels on your roof and sell energy into the electricity grid, this too is measured in kilowatt-hours. Depending on the feed-in tariff in your electricity supply contract, you might earn only five or ten cents for every kilowatt-hour of electrical energy you sell back to the grid.

If you have an electric vehicle (EV) or a home battery, the amount of electrical energy that you can fit into or extract from a battery is also measured in kWh, because we are still talking about an amount of electrical energy.

The tricky kW: a measure of power, not energy

Where it gets tricky is when we introduce the kilowatt (kW). The kilowatt is a measure not of *energy used* but of *power* – an instantaneous measure of how quickly energy is being generated or used in an instant or moment of time.

Forgetting the metric system of measurement for a moment, the imperial-system measure of power is the horsepower (hp), which may be easier to visualise than a kW. However, when using the metric system, the amount of power that a horse can exert in an instant while pulling a wagon is measured in kilowatts, not in horsepower.

The amount of power that an Olympic athlete can impart to a stationary bicycle in a moment of time is also measured in kilowatts. The power that a car's motor (petrol or electric) can output in an instant of time is measured in kilowatts.

The amount of electrical power that a light bulb might use in an instant of time is measured in the familiar watts (W). One thousand watts equal 1 kilowatt.

kilowatt × time = kilowatt-hours

The amount of electrical power (measured in kW) needed by a device such as a pool pump or light bulb, multiplied by the duration of time that the device is on for (measured in hours) gives us the electrical energy used (measured in kilowatt-hours).

Say, for example, a solar PV system on a roof at any instant in time produces electricity at the rate of 6 kilowatts. If the 6-kilowatt solar PV system can sustain that rate of power output over one full hour, it will produce 6 kilowatt-hours of electrical energy.

Here's another example. Let's say a 10-watt light bulb was left on for 24 hours. This would consume 240 watt-hours of electrical energy, which is equal to 0.24 kWh. If you are paying your supplier 30 cents for every kilowatt-hour of electrical energy, having that light on for twenty-four hours would cost seven cents.

A pool pump drawing 1 kilowatt (1 kW) of electrical power continuously over one hour will consume 1 kilowatt-hour (1 kWh) of electrical energy, which would cost you 30 cents, if again you are paying your supplier 30 cents for every kilowatt-hour of electrical energy.

Confused? Misuse of the nomenclature can confuse us all! The most common error is when people shorten 'kilowatt-hour' to 'kilowatt'. For example, when we hear a person say 'my solar system does 20 kilowatts' we're never sure if they mean their solar PV system has a maximum output capacity in any instant of time of 20 kW of electrical power, or if indeed their system is much *smaller* than that and only manage to produce 20 kilowatt-hours of electrical energy across one whole day.

Air conditioners: kW of electricity in, many more kW of cooling or heating out

With air conditioners there is even more potential for confusion. When you receive a quote to buy a '3-kW' air conditioner, what does that even mean?

Generally, the size of an air conditioner is quoted in the amount of cooling it can provide in a given instant of time (e.g. 3 kW of cooling output).

For a reverse-cycle air conditioner, there will be a similar figure that tells you its heating output capacity (e.g. a heating output capacity of 3.4 kW). This heating output figure can be understandable and useful because it means the reverse-cycle air conditioner will produce heat at a rate equivalent to 3.4 radiant electric heaters that have an output rating of 1000 watts each.

However, sometimes folks think these kW figures for air conditioners also refer to the amount of electrical energy an air conditioner consumes and therefore imagine an air con will be costly to run. This is way off the mark. In Chapter 5 (page 61) I describe how to divide an air conditioner's quoted heating output by the Coefficient of Performance (COP) to work out the required electrical power needed to run it. For example, assuming the above air conditioner has a COP of four, we can divide 3.4 kW of heat output by four to arrive at an electrical power requirement of only 0.85 kW, or 850 watts. This means this particular air con actually uses less electrical energy than a hair dryer.

SEE ALSO *Heating and cooling (pages 66–67)*

And finally, if that air conditioner draws 850 watts of power continuously for one hour, it would have consumed 850 watt-hours (0.85 kWh) of electrical energy. At a cost of 30 cents per kWh for electrical energy, running that air conditioner for one hour would cost 25.5 cents.

On to Part 2!

Now that I have finished setting the scene, let's move on to all the ways we can electrify our homes and get off gas. We will explore using heat pumps, induction cooking and solar PV panels, and improving our health and comfort by reducing draughts and enhancing insulation, windows and window coverings, all while shrinking our energy bills and environmental impacts.

Let's go.

Part 2
How to create your own efficient electric home

5.
Heating and cooling

For most Australian homes, the biggest energy cost is for heating and cooling your living spaces. This chapter discusses the cheapest, cleanest and healthiest ways to *actively* heat and cool (i.e. using heating and cooling appliances).

But don't forget about ways you can improve your home to *passively* keep it warm or cool, without using any heating and cooling appliances at all.[1]

You may live in a cooler Australian climate zone and be fortunate enough to have living areas with windows that face north and allow the winter sun to heat your home. A home's orientation to the sun is important to consider when buying a new home, choosing a home to rent, designing a new home or renovation, and when re-configuring living areas, windows and eaves.

It's vital to consider how to make the most of the winter sun – this is known as passive solar heating. Conversely, you also need to keep the summer sun from cooking your home.

But despite one's best passive efforts, most Australian households will at some point need to turn on a heater or cooler.

SEE ALSO *Do you live in a leaky bucket? (Chapter 8)*

The cheapest way to actively heat … is with an air conditioner!

In 2015 at the University of Melbourne, we were the first to work out how Australians could save a lot of money by using their reverse-cycle air conditioners for heating instead of burning gas. The savings added up to hundreds of dollars a year for a single household and billions of dollars across the country. I came to realise this was a very big cost-of-living win, but no one else was talking about it![2]

A reverse-cycle air conditioner can heat a home at around one-third the cost of burning gas and as little as one-fifth of the cost of using an electric-resistive heater, such as an oil column heater, fan heater or electric panel heater. As reported in 2015 in *The Age*, our research found that a large Melbourne household could save $658 a year by switching off the ducted gas heating and switching on the air cons.[3] The saving is probably even larger today.

I meet people all the time who are still unaware of this. One reason is that air conditioners have had a bad reputation.

Demonising air conditioners

Over the years, we've done a good job of demonising the use of air conditioners for cooling in summer. 'Don't use an air conditioner, it's expensive, you'll get a huge power bill', 'It will crash the electricity grid!', 'Use a fan, put a wet rag on the back of your neck, sleep in the nude!'

However, you may recognise that those comments pertain to using an air conditioner in summer. Indeed, if you can stay healthy and get by without using an air conditioner in the warmer months, I'm fine with that.

But what about winter?

A reverse-cycle air conditioner is an almost magical device that does two things: heat and cool. Amazing! I'm still amazed.

How often do you buy something for one purpose, and it turns out it's really good at doing something else? It's almost like you bought a car that is also very good at washing clothes.

However, reverse-cycle air conditioners aren't magic, they're actually heat pumps.

How heat pumps work

Heat pumps aren't magic, but they can be mysterious.[4] So how do they work?

You have at least one heat pump in your home already. It's your refrigerator. It uses its refrigerant cycle to move or 'pump' heat from one place to another, collecting heat from around your food and pumping it back into your kitchen. A fridge has to work almost continuously because heat from your kitchen is continuously trying to get through your refrigerator's insulation and into your food. But hopefully your fridge can keep up and your food and drinks remain nicely chilled.

A refrigerative air conditioner used for cooling works similarly to a refrigerator and is another type of heat pump. In summer, an air conditioner collects unwanted heat (not air) from within your home and pumps it outside. That's why in summer, if you ever walk near the outside unit of a functioning air conditioner, the air being blown around out there will be quite hot. The air conditioner has pumped heat out of your home and added it to the environment outdoors.

Note that in this process, no air moves from the inside of your home to outside, nor from the outside to inside. It's only the refrigerant fluid, travelling through small insulated pipes, that cycles between your living spaces and outside, carrying heat out of your home in summer.

If you want the detail, the five key parts of a heat pump refrigerative air conditioner are:

* a heat exchanger located inside your home (with air-circulating fan)
* a heat exchanger located outside your home (with air-circulating fan)
* the refrigerant that circulates from one heat exchanger to the other
* a compressor to pressurise the refrigerant (thereby making it hotter), and to push it around the cycle
* an expansion valve through which the refrigerant loses pressure and cools.

When an air conditioner is cooling a room, refrigerant flows through the expansion valve and gets very cold. This cold refrigerant next passes through the heat exchanger in your home's *inside* unit, where it collects heat from the room with the aid of the fan blowing air through the inside heat exchanger. This warms the refrigerant. The warmed refrigerant then goes through the compressor, raising the pressure and temperature of the refrigerant. It's now hot enough that when it next goes through the heat exchanger outside your home, the fan there blows outdoor air over the refrigerant, transferring heat to the outdoor environment and cooling the refrigerant. The refrigerant then flows through the expansion valve where it gets very cold, and the cycle begins again.

In winter, when a reverse cycle air conditioner is used for heating, refrigerant moves in the opposite direction. Refrigerant flows through the expansion valve and gets very cold, just as it did in summer. But this time the cold refrigerant passes through the heat exchanger in your outside unit and collects free renewable heat from the air outside your home, with the aid of the outside unit fan. The heat collected from the winter air outside your home warms the refrigerant.

The refrigerant then gets pushed along by the compressor, raising the pressure and temperature of the refrigerant, making it so hot that when the refrigerant next goes through the heat exchanger inside your home, the fan there blows room air over the refrigerant to transfer heat to your indoor living space. Giving up this heat cools the refrigerant. The refrigerant then flows through the expansion valve and the cycle begins again.

That's how a reverse-cycle air conditioner works in winter. Still people may wonder, 'If it's only 4 degrees Celsius outdoors, how can there be heat in the air?' The answer is there is always heat in our air, as long as the sun shines somewhere on our planet and our air remains above the temperature of absolute zero or minus 273 degrees Celsius.

It's correct to say that reverse-cycle air conditioners are actually another way to use solar energy. The sun heats our air and heat pumps collect free renewable heat from that air.

The temperature of the refrigerant circulating in a reverse-cycle air conditioner in heating mode might be as cold as minus 30 degrees Celsius, therefore that cold refrigerant will easily absorb heat from the air outside your home, even if it's only 4 degrees out there on a frosty morning. This means that heating with reverse-cycle air conditioners is a perfect fit for Australian homes. They are also used in far colder places, such as Canada and Finland.

Why a reverse-cycle air conditioner is so cheap to run

Why are reverse-cycle air conditioners the cheapest way to heat your home? Because as a heat pump, they collect *free* heat from the air outside your home. Yes, it requires some electricity to run the compressor and the fans, but this electricity use is leveraged several times during the process of transferring heat from outdoors to indoors.

Some people say heat pumps don't make heat; they simply move heat around. This is true, and importantly it costs a lot less to move heat around than it does to make it. This is like how it might cost you something to grow a tomato in your garden, but it won't cost you that much to then move that tomato from your garden to your kitchen.

In this way, a reverse-cycle air conditioner can be said to be operating at, for example, 400 per cent efficiency. A single unit of electrical energy (e.g. 1 kilowatt-hour) going into your air conditioner can result in 4 kilowatt-hours of heat coming out. That ratio of four to one (or whatever the exact ratio might be for the model of air conditioner you're thinking of) is called the

Coefficient of Performance (COP). For any air conditioner, you can find the COP listed among the device's technical specifications. My wife says I should express it this way: 'Buy one, get four.' She's never seen a deal that good.

Why are reverse-cycle air conditioners the cheapest way to heat your home? Because they collect *free* heat ...

A COP of four means that if you'd like heat coming into your lounge room at the rate of 2000 watts, you'll only have to put 500 watts of electrical power into your air conditioner.

How does this compare with a simple electric-resistive heater? An electric-resistive heater (e.g. oil column heater, fan heater or electric panel heater) can be 100 per cent efficient: 1 unit of electricity in equals 1 unit of heat out. While '100 per cent efficient' might sound good, it can't compete with the 400 per cent efficiency of an air conditioner.

A gas burner has even lower efficiency than any electric heater. The moment you buy gas, the first thing that happens is that you set it on fire. A good part of the heat available is immediately lost up the flue or chimney. Therefore, the heating efficiency of gas is always less than 100 per cent and can be even lower than 50 per cent for older gas heating systems.

It's very difficult for gas heating to compete with an air conditioner that may be at least five times more efficient. To compete, gas would have to be one-fifth the cost of electricity (on a $/MJ or $/kWh basis) and it's generally not. A common gas price I am seeing on Victorian bills these days may be $0.03/MJ, which is equivalent to $0.10/kWh. Electricity is more expensive than gas on a $/kWh basis, but it's not five times more expensive. I'm currently paying $0.28/kWh for electricity, which is only 2.8 times more expensive than gas priced at $0.03/MJ ($0.10/kWh).

During sunlight hours, Australian homes with solar PV panels on their roof can heat with an air conditioner for even less money. This may be the cheapest heat in the world. Sometimes it's actually free. How is the gas industry meant to compete with that?

Avoiding the sunk-cost fallacy

Here's about the saddest story I've found. I was at the home of a young couple, sitting in their lounge room beneath their split-system air conditioner. I asked them how they normally heat their house. They said they had just spent $4,000 to have their ducted gas heater replaced. But then they added, 'You know, we were lucky because we could use the split system above you there for heating for a few weeks before the furnace was replaced.'

The look on their faces wasn't great when I told them they could keep running the split system instead of their new gas heater if they wanted to save hundreds of dollars each winter.

Their story is not an unusual one. I hear it time and time again. It was $4,000 for the new gas heater down the drain, basically.

Never mind. What you've spent in the past on gas equipment, it's in the past! Avoid the 'sunk-cost fallacy', cut your losses and stop using the gas equipment. Even if it's only two years old or two weeks old, it doesn't matter, it's costing you. And since you likely already have or will soon be getting a reverse-cycle air conditioner for summer cooling, use this as your cheapest way to heat.

Get off the gas! Your now-redundant gas burning equipment can be sent off to the metal recyclers. The valuable metals won't be wasted.

Your gas heater could kill you but your air conditioner never will

Safety alert! To ensure poisonous carbon monoxide gas isn't entering your home, gas heaters must be checked by a qualified technician every two years.[5]

An air conditioner, on the other hand, won't kill you with poisonous gas.

Ditching gas-fired heating and using an air conditioner means more savings because there will be less need for frequent servicing. Although air conditioners (and gas heaters also, for that matter) have filters that need to be cleaned regularly, this is something that you can probably do yourself.

Beyond this, if an air conditioner has been operating fine for a period of four years or more with no professional servicing, it might be time to call in a technician to give the equipment a professional check and clean. I suggest you closely observe the professional air conditioner cleaning process. Learn what is found. Was there any mould hiding within? Is it particularly dirty? Discuss with the technician how to reduce the chances of mould growth or dirt build-up happening again. Ask how parts can be DIY cleaned.

Buying a new air conditioner? Should it be ducted or not?

You're reading this book and so perhaps you've decided, yes, you'll be getting your home off gas heating! Now you are in the market for a new reverse-cycle air conditioner. Should you get a ducted system or individual (ductless) split systems?

Generally, the most cost-effective way to set up a home with reverse-cycle air conditioning is to use individual split systems.

Why? Because ducted systems:

1. cost more to install
2. cost more to operate
3. restrict how you can use your rooms
4. cause you to heat more space than you need.

I explore each of these points below.

Ducted systems cost more

Ducted heating and cooling is also known as 'central' heating and cooling. Most commonly it means using ducts – either installed up in your home's roof space or down under your floor – to transfer heated or cooled air from a central heater/cooler to individual rooms.

Firstly, a ducted air conditioner is likely to cost more to install than an array of ductless individual split systems). (I describe the various types of 'split systems' on pages 74–77.)

Secondly, ducted heating and cooling systems can be less efficient and more costly to operate than ductless systems.[6] A lot of heat will be lost out through the ducts in winter and gained in summer. Why? Because in Australia:

✱ ducts are never well insulated
✱ ducts are commonly installed outside of the building's thermal envelope, either hanging near the ground beneath the floor or placed above the insulation up in the roof space
✱ duct installers might not do a perfect job given that they are often working in tight and uncomfortable spaces
✱ ducts are generally not inspected nor maintained and can be damaged by animals or humans crawling under the floor or in the roof space.

When we had ducted gas heating, the flow of warm air to the bathroom was never strong. I only went investigating a few years later and found that, beneath the house, a cat had chosen to sleep on the warm and cosy duct, crushing it.

After we had converted our home to heating with individual split systems, I pulled out the old gas-fired heating ducts. Most of them were in good shape, but I was sad to find in one place the installer had not joined two pieces together properly. This meant that for the past twenty years whenever we were heating with gas, we were constantly losing air and heat through an unsealed joint. Who knew?

More sensible building practice would see ducts being placed within the home's thermal envelope, as is commonly done in countries with more extreme climates. However, this arrangement must be established when the home is being built. It's not easy to retrofit.

Air must return to the return-air inlet

Thirdly, many home occupants are unaware that with a centrally ducted system, heated or cooled air must return to the return-air inlet. In other words, a centrally ducted system is meant to be operated as a closed-loop system.[7] What does this mean?

Once conditioned air comes into a room, it needs to find a clear pathway back to the return-air inlet, which is often located in a central corridor. In other words, when an air conditioning system is being used, bedroom or office doors must be left partially open to allow air to return to the return-air inlet. This means that zoning off sections of the home by closing interior doors cannot be successfully done with a centrally ducted system.

What this means in practice is that if you close the door of a bedroom when the central heating/cooling is on, you have choked off the system. The bedroom becomes over-pressured. The air that has come into the room can't find a way out other than to leave through a leaky bedroom window. On the other end of the system, there's not enough air returning to the return-air inlet. The system is now out of balance. To make up for the missing air, the system will be forced to draw air into the home from under the front door (for example). What was meant to be operating as an efficient closed-loop heating/cooling system is now an inefficient once-through system. Can you imagine the bills!

In many homes the occupants don't want to leave doors open. A person working in a home office may be communicating on Zoom and need to keep the door closed to reduce noise and enhance privacy. A child or shift-worker may need to sleep. A teenager may go into a bedroom, never to come out again. Therefore, doors will be closed. However, be aware that closing the door corrupts the operation of a centrally ducted system.

Ducted systems heat and cool more rooms than we need

Fourthly, with centrally ducted systems in Australia a household is likely to often heat more rooms than are being used. This is less of a problem overseas in severe cold or hot climates, where it's necessary that the entire home is heated or cooled at all times to stop, for example, pipes freezing in winter or to protect clothing and furnishings from mould damage during a humid summer. However, the climates most Australians live in generally aren't as demanding as that year-round.

Duct zoning – using valves that are placed in the ducts to prevent air flow coming down a leg of the ductwork – or even just closing off the heating outlets in a limited number of rooms, can save you conditioning some rooms that you aren't using. However, sometimes the entire system doesn't function optimally when sections are shut off like that. Talk to your installer or check the operating manual.

No filter means filthy ducts

And here's one last sad observation about the state of Australian ducted heating systems. There is meant to be a filter at the return-air inlet of ducted systems. A filter will keep dust, animal hair and other filth out of the ducting and the heat exchanger and blower downstream.

Sadly, more often than not, I find there is no filter! It may have been removed by a home occupant who grew weary of cleaning the filter. Or it may have been removed by a visiting heating technician who felt this was the best way to get air to circulate more strongly around the system.

If a ducted heating system has been run for a year with no filter, the ducts are going to be horribly contaminated with dirt and dust. Add dampness to the situation and this provides an excellent place for mould to grow. When I reach down into the duct at the return-air inlet that has not been protected by a filter, I grab a handful of dust and show it to the client. This insight alone is often enough for the client to immediately want to switch to heating with a ductless reverse-cycle air conditioner.

Intentionally removing the filter from a heating/cooling system in a short-term attempt to increase air flow and improve performance is akin to your car mechanic removing the filter from your car's petrol engine so that you can drive down the Hume Highway a little bit faster. We'd never do such a damaging thing to our cars, but it's a quick fix I see all the time with ducted heating systems.

I've concluded that centrally ducted systems operate inefficiently in Australia and often don't suit the way we want to use our homes, whereas using ductless air conditioners allows the home's occupants to optimally control the conditions in each individual living area and not waste energy and money. Doors can be closed, and/or other means used (such as hanging a sheet or Japanese-style fabric 'noren' in a connecting hallway or stairway) to segregate spaces and heat or cool individual living spaces only as needed. This is a much more cost-effective way of heating or cooling your home.

SEE ALSO *page 89 for more on the noren*

Given they can have big advantages over centrally ducted systems, I'll now describe the different types of ductless air conditioners.

Types of ductless reverse-cycle air conditioners

It's a phrase that I haven't heard anyone else use, but I refer to air conditioning systems that involve no ductwork as 'ductless' systems. They come in many shapes, sizes, designs and configurations. If you will be buying an air conditioner, take time to investigate everything that is available.

Wall-mounted split systems

Many of us are familiar with the common up-on-the-wall equipment referred to as a split system. There is one piece of equipment indoors and one outdoors.

This is a ductless type of air conditioner because there are no air ducts involved. Indoors, air is drawn into the top of the split system, filtered, heated or cooled, and then blown out the front. No air leaves nor enters the home through any ductwork.

However, refrigerant does circulate between the inside and outside units through small-diameter insulated piping carrying heat into or out of the home. Again, no air travels between the inside and outside parts of the split system.

Through-the-wall/window air conditioners

Prior to the invention of two-part split systems, common air conditioners came in the form of a single-piece box-shaped machine that would sit in a window or be placed through a wall, with the electrical connection being a simple wall plug.

These 'non-split' window or wall air conditioners can still be bought today. They can be used for cooling only or both heating and cooling (aka reverse-cycle). Modern units will be more energy efficient than older models. I've been in the homes of clients who were happy using this style of air conditioner for heating and cooling.

They sound simple, so why are they less common than split systems? One reason is that when installing a new air conditioner in an existing home (that may have been using gas heating), split systems require only a small hole to be punched through the wall, through which the refrigerant lines, condensate drain line and electrical connections pass. In comparison, a non-split wall air conditioner would need brick and/or carpentry work done to create a large hole through the wall for the entire machine to sit in.

'Floor-mounted' air conditioners

The common split systems mounted up on the wall can easily cool a room because cool air is heavy and falls toward the ground, where we may be lying in bed or sitting in a chair.

However, in heating mode, more effective heating would be achieved if your air conditioner was mounted down near the floor, since hot air rises. Warm air would more directly heat the lower part of the room where the people actually are. Your heater is of little use if the hot air starts and stays up near the ceiling, while your feet remain cold. This can be a bigger problem in rooms with high ceilings.

Fortunately, to address this issue, floor-mounted reverse-cycle air conditioners do exist. They are also known as 'floor-standing' or 'console' units. Ask your supplier about them. They are common in New Zealand and Tasmania because folks there know an air conditioner/heat pump is essential for winter heating rather than summer cooling. Though it may be floor-mounted, it's still an air conditioner: it heats, cools and filters the air just like an up-on-the-wall model. It's just packaged differently.

Bulkhead and cassette style

One reason people are attracted to ducted systems is because they are visually discreet. Some people have strong adverse reactions to air-conditioning equipment that can be seen within a home. 'I won't have that ugly thing up on my wall!'

As a result, the designers of ductless air conditioners have thought up ways to make their equipment less noticeable. 'Bulkhead units' can be hidden in bulkheads (a dropped part of a ceiling that has been sectioned off or enclosed) revealing only two grilles, one where air is drawn into the air conditioner and the other where air is blown back into the room. There are no air ducts with this style.

Cassette units are flush-mounted in the ceiling. You may have seen these in an office or restaurant, but you can get one for your home. They'd make most sense in the ceiling of a large open-plan living space. Again, there are no air ducts involved. Air is simply drawn into one part of the cassette, filtered, either heated or cooled, and then blown back out into the room via the cassette outlet.

Bulkhead and cassette units will be more expensive to install than the more common up-on-the-wall split systems.

Multi-head (aka multi-split) air conditioners

I've described how ductless air conditioners have one unit inside the home and one outside, connected by refrigerant piping. We have two such air conditioners at our home. My son's home has five split systems. Some homes have as many as seven or eight individual air conditioners. Each of these may come with both an inside and outside unit.

People often ask, 'Is there a way not to have so many units on the outside of my home?' Yes, there is a way. It's referred to as a multi-head or multi-split system, meaning you can have one outdoor unit sending refrigerant off to four or even more inside units. This can be useful not only for aesthetic reasons, but also in cases such as inner-city homes where it's not possible to find a place to put multiple outside units.[8]

Be aware, however, that the cost to buy and install a multi-head air conditioner is often higher than for individual units. This seems strange because you are buying less equipment. However, since not as many multi-heads are sold as individual split systems, the prices may be more expensive, and the installation of multi-heads can be more complicated because of the need to send refrigerant piping off in more than one direction.

If you are thinking of buying an air conditioner, ask your installer about everything they have to sell, so that you are aware of *all* possible options.

What size air conditioner do you need?

When you get quotes for a new air conditioner, the supplier will recommend a suitable size. Generally, they'll err on the side of giving you a unit larger than what you might need. Of course, they don't want to get complaints later that you think the system they installed is too small, not powerful enough or takes too long to heat or cool a room.

With modern air conditioners, having one that is a bit over-sized isn't a big problem because even if you never use its full capacity, it will run efficiently at low output.

If you want to have a go at sizing an air conditioner yourself, check out the website fairair.com.au.

Is an electric-resistive space heater ever a good idea?

Previously I mentioned that a simple electric-resistive space heater, such as an oil-column heater, fan heater or radiant-electric panel heater, will heat at five times the cost of an air conditioner. Does this mean that you should never use an electric-resistive heater? No.

In our house, as well as having two reverse-cycle air conditioners, we also have a small portable electric-resistive fan heater in the bathroom, which we use during the coldest weeks of the year to warm this small space during morning showers.

Upstairs we have two rooms that were formerly our children's bedrooms but are now home offices. We haven't yet invested in permanently mounted air conditioners for these rooms because this would cost around $2,000 dollars each. Rather, we continue to get by with small electric-resistive heaters under our desks that are used for a limited number of hours each winter. I track our electricity use and so I know they aren't costing that much. If we thought that cooling was going to be needed in these rooms, or that we would be using the electric-resistive heaters more in future, it would be worth us investing in additional reverse-cycle air conditioners.

SEE ALSO *Minimising your electricity use and costs (page 169) for tips on how to track your electricity use*

What about hydronic heating?

Hydronic heating refers to heating by pumping heated water around the home, either through wall radiators or through pipes within a concrete slab floor (also called hydronic underfloor heating). The water can be warmed either by burning gas or by using an electrically driven heat pump.

As the heat produced by hydronic heating is transferred to a person's body as radiant heat (as opposed to heat being moved by blown air), people often refer to hydronic heating as 'lovely'. And it is! It's especially lovely in homes that are draughty and poorly insulated because at least you can sit right next to the radiator and feel warm.

I'm familiar with many home owners who have gas-fired hydronic heating systems. They'd like to stop burning gas but would also like to retain the hydronic way of heating. This goal can be met by replacing the gas furnace with a heat pump. Europe is making this switch in a big way as they try to get off Russian and other sources of fossil gas.

The problem is that retrofitting a new heat pump to an existing hydronic system comes with a big price tag. So, in that situation, I always ask if instead of heating with a hydronic system, the home can be totally or at least partly heated with reverse-cycle air conditioners. After all, with our Earth overheating, you're probably going to need summer cooling anyway, which is not something that a hydronic heating system commonly does in a satisfactory way.

I've also met with people building new homes who are considering hydronic heating. I point out to them that if they build a home with a very good thermal envelope to the 7-star standard or even better, their new home will perform well and is not going to need a lot of heating. In other words, in a newly built home, investing in a hydronic system is overkill. With a new well-performing home, all that should be needed in winter or summer is a bit of reverse-cycle air conditioning.[9]

SEE ALSO *Do you live in a leaky bucket? (page 116) for more on just how easy it can be to heat a 7-star home*

A family member bought a house that came with a fairly old gas-fired hydronic heating system, along with three reverse-cycle air conditioners. We added two more reverse-cycle air conditioners for two bedrooms that required summer cooling. We removed the hydronic heating system, the gas boiler, radiant panels and pipework, and had these all thoughtfully recycled. The family members were glad to see the hydronic panels removed to free up wall space. The house is now gas-free, and we avoided the $15,000 cost of adding a heat pump to the hydronic system.

Getting the most out of heating with your air conditioner

Some households, on switching from gas-fired heating to a reverse-cycle air conditioner, are happy! That was the case in our home when we switched from using ducted-gas heating to using two up-on-the-wall split systems, one in the main living area and one in the master bedroom. We knew we were doing the greener thing by no longer burning gas. As well, we realised we were saving hundreds of dollars each winter.

However, some other households complain that the air conditioner, when used for heating, doesn't make them feel as warm as they used to feel. My first question in this situation is: 'Is the filter clean?' For best air conditioner performance, filters should be cleaned whenever they appear dirty. My next suggestion would be to check that the air conditioner itself is in good working order.

If the filter is clean and the air conditioner is in good working order, then I encourage householders to study the air conditioner's operating manual and experiment with different temperature and fan settings. Avoid settings such as AUTO unless you are confident you understand what AUTO actually does.

Don't be concerned if you have been told setting a temperature at, for example, 24 degrees Celsius, is too high, if that is what it takes to make you comfortable. Keep in mind that although 24 degrees may seem like a high number, it's being measured up at the air conditioner itself, not across the room where you might be seated. Use a cheap portable thermometer to measure the temperature where you are sitting and compare that with the air con temperature set point.

Even after all that, some people new to heating with an air conditioner are still uncomfortable. Recent research at RMIT has confirmed why.[10] Often our homes are draughty and poorly insulated and have a lot of single-glazed glass. RMIT's computer simulations reveal that it's just not possible for a reverse-cycle air conditioner to make such a room feel super comfortable. The powerful old gas heater, blasting out higher-temperature heat than a reverse-cycle air conditioner, would have previously been compensating for the poor performance of a home's thermal envelope.

My recommendation in this case is to examine the easiest ways to improve a room's thermal envelope. Better roof space or wall insulation may be called for, along with draught-proofing and better interior window coverings that can be drawn closed on the coldest winter nights.

SEE ALSO *Do you live in a leaky bucket? (page 116)*

Heating alternatives

Each year as you head into winter, there may be things you can do to delay turning on the heater.

One idea is to dress appropriately. If you've not heard of an 'Oodie', check these out. They are a wearable, fleece-lined blanket and they come in all shapes and sizes. On many cold Melbourne mornings when I jump out of bed, putting UGG boots on my feet and an Oodie over my head, turning the heat on is unnecessary.

Electric throw rugs and electric blankets make good personal use of electric-resistive heating. The concept here is to heat the person instead of the whole building.

Still, at some point during an Australian winter it can be necessary to heat a living space, or more particularly a bedroom, in order to ensure that the ceilings, walls, rugs, bedding, drapes and clothing don't become so damp that mould sets in.

SEE ALSO *Insulation: a priority (page 144) for more*

HEAT THE PERSON...

NOT THE SPACE

Summer cooling

If you've been able to avoid having air conditioning equipment at your home so far, that's great. But don't punish yourself in the future. Our Earth is overheating. Many premature deaths occur during heat waves, especially among the very elderly and those with health challenges.

Certainly, it makes sense to pursue passive ways of keeping your home cool: encourage plants to grow around the home, protect windows outside and inside from the effects of the direct sun and heat reflected from surfaces, open windows to try to catch a cross-breeze, close up the home if it's going to be a hot day.

SEE ALSO *Windows and window coverings (page 152) for more on protecting windows*

You can also use fans to get the air moving and assist your body's natural cooling mechanisms. Ceiling fans are useful, but I often prefer a simple floor-mounted pedestal fan so that air can blow directly at me. One advantage of using a pedestal fan over a ceiling fan is it might allow the hot layer of air up near your ceiling to remain undisturbed and not spread throughout the room. (Although we have a number of pedestal fans, I've never actually paid for one. For some reason in my neighbourhood, people are always leaving serviceable pedestal fans out on their nature strip!)

Also, try the FAN ONLY mode on your air conditioner. That will also do a good job to get the air moving. And compared with turning to COOL mode on the air con, fans don't cost that much to run.

Additionally, I'll often use a damp cooling cloth around my neck to help the fan. Cooling the person instead of cooling the entire living space will reduce your cooling costs.

But despite your best efforts, there is likely to come a hot summer when air conditioning of one type or another becomes essential. With more of us working from home, well, I find it difficult to think and do serious office work when the indoor temperature exceeds 27 degrees Celsius, even if I have a fan blowing in my face and a wet rag on my head.

Two methods of cooling commonly used in Australian homes are refrigerative cooling (such as with a reverse-cycle air conditioner) and evaporative cooling. Which is better?

Refrigerative air conditioning versus evaporative cooling

I've already described how refrigerative heating and cooling works. Electricity is used to power fans and a compressor, which push air and refrigerant around. This sort of heating and cooling works best in a draught-proofed and insulated home. If you have a ducted system, it should be operating in a closed-loop, which means that air comes into a room and then returns to the return-air inlet, and then goes around again without leaking air to the outdoors or sucking air in from the outdoors.

Evaporative cooling is very different from refrigerative cooling. This method uses the evaporation of water to cool a stream of fresh air that is brought into the home via ducts in the roof space and through air outlets that are cut into the ceiling plaster. When using evaporative cooling, some windows need to be open. This means that, unlike refrigerative cooling, evaporative cooling is not a closed-loop system: it's 'once through'. The humidified and cooled air comes into the home, picks up some heat from the rooms and immediately exits via the windows, never to be seen again.

Evaporative cooling works best where water is cheap, where the air is not too humid and where it doesn't really get that hot. This is because evaporative cooling can only drop the temperature of your home a certain number of degrees, which depends on how humid the air outside your home is to begin with. Unlike with refrigerative cooling, you can't just dial up 25 degrees; you get what you get with evaporative cooling. Under humid outdoor conditions, you might not get much cooling at all.

With evaporative cooling, it doesn't matter if your home is draughty and leaks like a sieve because you want the humidified air to immediately leave anyway – out through any open window or other gap in the building fabric. In comparison, the best refrigerative cooling results will be achieved in homes that have been draught-proofed.

In places like Melbourne, evaporative cooling was widely used twenty years ago. In my view, it's still too widely used today. It's a technology for a bygone era and is past its use-by date. Refrigerative cooling is far more capable and efficient than it was twenty years ago when evaporative cooling held sway. Additionally, many households these days find that the humidified feel of evaporative cooling is not what they're looking for in summer cooling.[11]

Nevertheless, in places where it's never very humid and where water is cheaper, or in cases where a home is hopelessly draughty and/or already equipped with evaporative cooling, it may be a better option than using refrigerative cooling in summer.

But here's something to think about. You can't heat your home in winter with evaporative cooling, whereas you can heat your home with a reverse-cycle air conditioner.

It concerns me greatly that new homes are built in Victoria fitted with two independent space-conditioning systems – outdated evaporative cooling and ducted gas heating – when a single modern refrigerative system easily does both heating and cooling. Thankfully, in Victoria this will change as new laws restrict the use of gas in new-build homes. [12] A knock-on effect of the gas ban is that we'll stop seeing evaporative cooling.

In existing homes with evaporative cooling, there are other problems. It can cost you more in winter than in summer! How?

Those massive holes cut through the ceiling plaster allow a huge amount of heat to be lost. I've had clients who felt they had cut their heating bill in half simply by fixing airtight covers to the evaporative cooling outlets. So seasonally (if not permanently), seal off those evaporative cooling holes in your ceiling.

Plastic covers are available to go over the evaporative cooling outlets. (Note that any louvres that come as part of the evaporative cooling outlets for directing air flow can be closed somewhat, but don't provide an airtight seal.) If outlet covers aren't readily available in your area, one quick fix we did at my daughter's home was as follows. We pulled out the louvered sections (usually four) individually, gave them a clean, wrapped them with cling wrap and popped them back in. Draught-sealing on the cheap!

Also for homes with evaporative cooling, those outlet holes are places where, clearly, your ceiling isn't insulated. Furthermore, the 'spaghetti' of ducting in the roof space often gets in the way of insulation installers doing the best job.

My recommendation for many places in Australia is to heat and cool your home, if needed, with a reverse-cycle air conditioner. If your home has evaporative cooling, have a think about getting rid of it, plastering up the holes in your ceiling, and upgrading the insulation in the roof space once the now-redundant ducts are pushed out of the way or completely removed.

What about portable air conditioners?

When cooling with a fan isn't enough, and there's no permanently installed air conditioning at your home, your thoughts may turn to using a portable air conditioner. These come as evaporative or as refrigerative types. Let's examine how they work.

Portable evaporative cooling

If you are in a dry climate, a portable evaporative cooler may help. I've heard them referred to as 'swamp coolers'. I remember using one in the 1980s in Melbourne when my wife was pregnant and we were desperate! It helped a little, but if your local climate is already humid, evaporative cooling sure can turn your lounge room into a swamp, and you might not be that comfortable.

Portable refrigerative 'single-pipe'

This type of air conditioner is widely purchased in Australia but, given how inefficient they are, I'm not sure if they are widely used. People may use them once and then have second thoughts about ever using them again.

As a single-pipe portable air con is blowing cool dehumidified air in your face, it will be busy blowing hot air out the pipe situated in a temporarily-rigged but sealed opening at a window. However, here's what people don't think about. If air is being blown out of the house, somewhere else air is also rapidly being drawn in.

Stealthily, a portable single-pipe air con will be heating up every other room in your house as hot and possibly humid air sneaks in through gaps and cracks around windows and elsewhere. It has to, in order to balance the flow of air that it is blowing out of the house. In this way these portables are very inefficient. They can cool a small space where you are sitting but at the same time heat up your entire house.[13]

For this reason, I see this type of air conditioning as being for emergency use only! A friend once organised a dinner party. The evening was forecast to be very hot. So, of course, his split system broke down the day before. He hastily arranged for a portable air con to get his guests through the evening. And sure enough, as the air con cooled the immediate area where we were dining, it was remarkable how hot air was drawn in and heated up every other room in his house.

I've not seen the more efficient 'double-pipe' portable air conditioners for sale in Australia. This design is far more efficient/effective than the single-pipe design because they work more as a 'closed-loop' and avoid drawing hot air into your house. I have seen instances where people have converted their single-pipe unit into a makeshift double-pipe unit.

In-the-window (but not portable)

Lastly, if you have a suitable window, you may be able to safely place a single-piece, non-split box air conditioner in there. We do this seasonally upstairs in a spare bedroom if we are having guests sleeping there or a family member is isolating from a summertime case of COVID-19. As I mentioned before, these are available as reverse-cycle for heating and cooling. Though the basic design has been around since about the 1960s, modern versions can be reasonably efficient, although not as efficient as the best split systems. They are certainly more efficient than a single-pipe portable because by sitting in the window, they essentially act as a double-pipe model.

What's the disadvantage versus a portable? They are heavy to move around. Be sure to install a support to keep it from falling out of the window and onto some passer-by walking beneath.

Don't forget FAN ONLY mode

Modern reverse-cycle air conditioners have a FAN ONLY mode. It doesn't use much electricity and, like any other fan, can help keep you cool. Ceiling fans can be excellent, but an air conditioner's FAN ONLY mode can perform a similar function; in fact, after we installed reverse-cycle air conditioners, we removed a ceiling fan from a low ceiling. If you're tossing up between buying an air conditioner or a ceiling fan, I suggest you'll want an air conditioner for the hotter times to come. By using the FAN ONLY mode, you might never need to install a ceiling fan.

Zoning living spaces

One way to reduce home heating and cooling costs is to heat and cool only those parts of the home that are being used at any point in time. This can be referred to as 'zoning'.

A centrally ducted heating and cooling system may have been installed with zoning valves in the ductwork that allows air flow to only some parts of the home. Zoning can be even easier with ductless air conditioners. With these, the householder turns on only the air conditioners that service the parts of the home being used at any time.

Installing interior doors can allow zoning (but as described previously, aren't recommended if central heating/cooling is used). Of course, this runs counter to the large 'open plan' living spaces that have become popular. Homes with large atriums and voids that extend from the ground floor to upper floors can make zoning impossible and it will be challenging to heat and cool to comfortable levels.

In our home, we have a simple stairway leading to the upstairs. In the past with the heating system on, I would sit on the stairs and feel cold air cascading down my back as hot air rushed to the upper floor. There was no permanently installed door to close off our stairway, so how to avoid heating the whole upstairs when no one was up there?

The answer was the 'noren'. These days in winter we hang a simple decorative sheet of cloth to zone the upstairs off from the downstairs. However, we don't call it a sheet, we use the Japanese term 'noren' because that sounds fancier. The noren allows our downstairs reverse-cycle air conditioner to warm up the downstairs space quickly, which is where we spend most of our time. A thin sheet of cloth is all that is required to prevent hot air from moving downstairs to upstairs. Because our house is not draughty, the noren hangs perfectly still unless I have left a window open upstairs. I have measured the room temperature at 20 degrees Celsius on one side of the noren and as cold as 14 degrees on the other. That's a dramatic difference created with just a simple sheet of cloth.

Here is another idea. Sometimes it is difficult to improve a draughty room such as a toilet, bathroom or laundry. If the room has a door, cheap (less than $10) twin draught-stoppers, or what I call 'double door-snakes' can be bought and simply slid beneath the door. Then close the door to isolate or zone off these draughty rooms from the rest of the home.

To further enhance the zoning effect, the full perimeter of an interior door can be unobtrusively draught-proofed using a product such as the 'EMV Slimline Door Perimeter Seal'.

6.
Heat your water the cheapest way

Having considered how to heat and cool your living spaces, the next major use of energy in Australian homes is generally for water heating. This chapter describes ways to save money, water, energy and reduce the use of fossil fuels, thereby reducing greenhouse gas emissions as you heat your water.

How much hot water do you need?

Let me first ask, why would anyone need a dedicated water heater? Firstly, do you need a water heater for washing clothes? Perhaps not. Clothes can often be washed in cold water. We bought a new clothes washer that connects only to the cold-water tap and can heat its own water. This means the water isn't heated via our efficient hot-water heat pump, but rather with a resistive-electric heating element in the washing machine itself.

Secondly, does a home need a dedicated water heater for dishwashing? Our dishwasher doesn't heat its own water; a dedicated water heater is necessary for a dishwasher like that. Many people wash dishes by hand. That hot water could come from a dedicated water heater or even from an electric kettle.

A third reason you need hot water is for washing your body. At the sink perhaps, or in the shower. A child might have a bath in a tub.

A fourth reason is hydrotherapy – a steamy shower can help to clear sinuses. Others might like to relax in a bath from time to time.

So that's probably where most of us use hot water. The point I'm making is that you might not need to use that much.

A shower head fit for a 5-star hotel

 A brand of shower head I like is the Methven Kiri Satinjet. Why? The people who invented this actually thought about the problem of water and energy conservation. Rather than having a whole bunch of individual jets that individually spray out water, the jets on a Methven Kiri Satinjet are what I call 'double-impinging' jets. Water comes out of one of these double jets and immediately collides with water coming from the second jet. This aerates the water and provides a nice-feeling shower but with low water use.

One weekend we stayed at the Sofitel hotel in Melbourne. When I stay at a 5-star hotel, the first thing I do is to look at the shower head. Sure enough, the Sofitel featured the Methven Kiri Satinjet. Good enough for the Sofitel, good enough for me.

Saving money at shower time

Next time you have a shower, grab a bucket and check how much water your shower head is passing. Make sure you aren't using water at a rate of more than 9 litres a minute. This is the flow rate to expect from a standard 3-star shower head. If you want to do better, there are 4-star shower heads that use as little as 4.5 litres a minute.

A cheaper yet often acceptable alternative to buying a new shower head is to buy, for only a few dollars, a plastic flow regulating orifice disk. They come three or four to a packet and allow water flows of either 6 or 9 litres a minute. You will need to take the shower head piping apart to fit the flow restrictor in.

The cheapest way to heat water

What's the cheapest way to heat water these days? My general recommendation is to consider heating your water with a hot-water heat pump.[1] Operating similarly to a reverse-cycle air conditioner in heating mode, hot-water heat pumps extract free renewable heat from the air outside your home. The nice thing about free heat is that it's free. A nice thing about renewable heat is that it's renewable. There's heat in the air outside your home, so why not use it?

Yes, a hot-water heat pump does require some electricity to operate its fan and compressor. However, for every unit of electricity you apply to a hot-water heat pump, you can get three or four units of hot water coming out. In this way, a hot-water heat pump can be said to be around 400 per cent efficient.

Therefore, using a hot-water heat pump can be the cheapest way to heat your water, operating at around one-third of the cost of burning gas or around one-fourth of the cost of using a traditional electric-resistive hot water service.

But wait, there's more! Hot-water heat pumps can heat water especially cheaply if you happen to also have solar PV panels on your roof. Using a timer, you can set your heat pump to come on toward the middle of the day when your solar PV panels are likely to be generating cheap electricity. It will then run for two or three hours to ensure all the water in the tank is hot, while consuming less electricity than a hair dryer.

In this way, a hot-water heat pump acts as a 'thermal battery', collecting and storing heat one day to be used the next. A home with a hot-water heat pump and solar PV panels on the roof will enjoy perhaps the cheapest hot water in the world, using practically free and entirely renewable energy.

Don't be left standing cold, wet and naked!

How old is your water heater? There may be a date of manufacture printed on a label stuck to the side of your hot water service or behind a removable panel. If your water heater is more than ten years old, it's a good idea to start looking into replacement options. Water heaters have a reputation of failing on Christmas Eve, which is not a great time to begin your heat pump research. Imagine you are in the shower and the water fails to heat up. It's difficult to think through a good investment decision when you are cold, wet and naked! Or even worse – when your partner has been left standing cold, wet and naked.

There are various rebates and incentives available in states and territories across Australia for replacing electric-resistive and gas hot water systems with a hot-water heat pump. These rebates and incentives won't be around forever, meaning that right now could be a good time to switch to one.

Hot-water heat pump brands, refrigerants and capabilities

Hot-water heat pumps were a niche product in Australia over the past twenty years, but no longer. They've gone mainstream. There are many brands available, ranging from those that perform extremely well to those that I would not recommend you buy.

One way to grade this wide range of heat pumps is by the refrigerant used. The most capable hot-water heat pumps use carbon dioxide as refrigerant, which goes by the refrigerant code of R744. Next are the hot-water heat pumps that use propane refrigerant (R290), followed by those that use the refrigerant chemical 1,1,1,2 –tetrafluoroethane (R134a).

So, which to buy? Which would fit best with your home? I liken buying a hot-water heat pump to buying a car.[2] There are many things to consider. Unfortunately, you're not going to be able to go to a heat pump shop and take one around to your home for a test drive. The best you may be able to do is to check reviews online, ask suppliers about where you can see and listen to one operating, or perhaps ask around at online forums such as the public Facebook group MEEH and at other community groups.

The consumer product reviewers CHOICE have on their website free technical specifications of some hot-water heat pump brands available in Australia.[3] Although not serving the entire country, the business Want a Heat Pump (as one example) offers a variety of brands, illustrating the range that is available.

Hot-water heat pumps: a checklist

There is a long list of things to think about and compare when buying a hot-water heat pump, including:

* Which **refrigerant** is used?
* What is the **global warming potential** of the refrigerant if it is leaked or released to the atmosphere?
* What is the heat pump's **energy efficiency** across the full operating range of ambient (outdoor) temperatures and what does this mean for comparative operating costs?
* What is the **minimum** possible ambient operating **temperature**?
* Is an **electric-resistive element** included to deal with low ambient temperatures? How often will it need to be used throughout the year?
* How quickly can it heat or **re-heat**?
* How many **renewable energy credits** (small-scale technology certificates or STCs) will be applied?
* How **noisy** is it? (Note that the decibel scale is not linear. Sound pressure level doubles for each 3-decibel increase, meaning that a difference between 38 decibels and 44 decibels is a quadrupling of noise.)
* Is the device arranged as a two-piece **split system** or an **all-in-one** configuration?
* Will a separate **electrical connection** be required? What amperage will this need to be?
* What **tank size** do you need: 170 litres, 400 litres or something in between?
* What is the **physical size**, weight and footprint of the device? Will it fit down the side of the home or up the stairs of a unit?

* **Where** can it be installed? On the balcony or on the roof? In a garage? Should the tank of a split system be placed indoors to keep it out of the weather?
* Will the **tank material** be stainless steel or enamelled steel with corrosion protection (sacrificial anode)?
* What are the **maintenance** requirements, such as the need to replace a sacrificial corrosion anode?
* Does it come with **a timer** so you can coordinate with solar PV-generated electricity? How easy and flexible is the timer to use?
* What are the **warranties** for the various pieces of equipment?
* How **reliable** will it be? How quickly will it be fixed if there is a problem and how many years will it last? What are the likely mechanisms of failure?
* What is the experience and **reputation** of the equipment supplier and the installer, and will they both be available and work well together should a problem develop?
* How much **local content** is there: is any of it made or assembled in Australia?
* What **rebates and credits** can be applied in your state or territory and in your particular circumstances? Are rebates means tested? At what level of income?
* What is the installed purchase **price** – after all eligible rebates and credits are applied?
* What is the **delivery time** – when can you get one? What can you do in the meantime if the current hot water service has failed?

What size hot water tank will you need?

I will try tackling just one question from the list above. What size tank will you need? The different brands often offer a larger and a smaller tank. Some brands will have a full range of sizes.

Previously I described how you should aim to have your shower head use no more than 9 litres a minute of water. Let's assume 5 litres of this will be hot water and the rest cold. Were you to stand around in the shower for 10 minutes, which is a long time for an adult, you would use 50 litres of hot water. Tanks come in a range of sizes from around 170 litres up to 400 litres. You can see that a smaller-sized tank would be more than adequate for one or two people. Remember that if you don't constrain your hot-water heat pump with a timer, it will reheat as soon as it senses cold water coming up to the level of the thermostat.

However, if your home has much higher hot water use and you intend to use a timer to constrain your heat pump to running only in the middle of the day, to make best use of your solar PV electricity you'll want a larger tank.

Electric water heating options for small spaces

In many Australian homes, water is heated with a small tankless (often referred to by the marketing terms 'instantaneous' or 'continuous flow') gas burner. These are attached to a wall and do not take up much space. On the other hand, a hot-water heat pump involves a water storage tank that indeed has a significant footprint. These can be challenging to position in some homes that don't have a utility or equipment space running down one side or out the back. Heat pumps extract heat from the air, so they are best installed either outdoors or in large, enclosed spaces, such as in a garage or basement.

Some heat pumps have an all-in-one configuration, with the heat pump part attached right on top of the water storage tank. Others are split systems, where the heat pump is separate from the tank. The split system arrangement can offer some flexibility when it comes to finding a place for the equipment.

The installers of hot-water heat pumps can be creative in finding a location for a heat pump and tank, so I suggest you work with them to learn what is possible at your home.

If you find there is no place for a hot-water heat pump at your home, there are other electrical water heating options with small footprints.[4] If a household doesn't use that much hot water, choosing a small electric-resistive tank may allow for other siting options.

There are also tankless electric-resistive water heaters that don't take up much space at all. I recall having a couple of these when I lived in an apartment in Germany in the 1980s. However, these may require a 'three-phase' electrical connection (triple the electricity supply capacity of a normal or ordinary single-phase connection) to ensure adequate hot water on the coldest winter day. This type of tankless water heater stores no hot water and therefore can't act as a thermal battery, meaning you can't take as much advantage of cheap solar electricity generated on your roof as you can with a hot-water heat pump.

SEE ALSO *Bye-bye, gas grid: cooking without gas (page 103) for more on three-phase power*

Why not use a solar-thermal hot water system on your roof?

Solar-thermal hot water systems have been widely used around Australia for more than twenty years. Back then they were about the only renewable energy thing we all could do at our homes, so they became popular. If you mentioned a solar system on your roof in those days, you would have been talking about making hot water and not electricity. Government regulations mandated the use of solar-thermal hot water for some new homes.

Solar-thermal equipment includes a panel on the roof to collect the sun's heat, and an electric or gas-fired booster to ensure hot water is still available even if the sun is not strong for a few days in a row.

One problem with solar-thermal hot water systems is that the solar part may not do much at all outside of the sunniest summer days. Another is that when something associated with the solar-thermal part fails, such as a circulating pump, a temperature sensor or a non-return valve, your household carries on unaware of this failure, as hot water continues to be supplied thanks to the electric or gas booster.[5]

So if you have a solar-thermal hot water service at your home, try to work out if the solar part is actually doing anything for you. You can do this by feeling the pipes to see if any of them are warm. The next question is: are the pipes warm because the solar-thermal part is functioning or only because the booster is heating the water? If all pipes are cold, this might mean the solar-thermal part is not doing much. I may have a degree in chemical engineering, but when visiting a home I am often baffled trying to diagnose what may have gone wrong with a solar-thermal hot water system.

Because of the cost and complexities associated with solar-thermal hot water, heat pumps are taking their place. Even companies that used to sell solar-thermal hot water are now selling heat pumps instead. Hot-water heat pumps can extract free renewable heat out of the air all throughout the year, day or night. With a hot-water heat pump there is nothing to go on your roof. This allows you to reserve that valuable real estate for solar PV electricity-generating panels.

Waiting for hot water to reach the tap

We would all like the hot water to arrive at our tap as quickly as possible. It's wasteful to see litres and litres of cold water going down the drain while we wait for the hot water to arrive.

The tankless gas-fired hot water services are often marketed as being instantaneous. This has been clever marketing because, of all hot water systems, they can be the least instantaneous, with users often complaining of the time it takes before it fires up and actually begins to heat water.

There is nothing more instantaneous than having a tank of hot water sitting waiting for you. This is what you get with a hot-water heat pump.

Still, with any hot water system there will be a delay in receiving hot water unless the hot water storage tank is very close to the tap. In our home that is the situation with our shower. When we turn on the hot water tap, it's only a second or two before the hot water arrives. Whereas at our kitchen sink, 2 litres of cold water must be purged from the water pipe before the hot water gets there. We use this to water plants.

When considering where to locate a new hot water service, think about how long the connecting pipes will be and how much time it will take before the hot water arrives at the tap.

Avoid using an energy-hungry ring-main recirculating hot water system. While these provide the convenience and water savings of having hot water always available immediately from all of the hot water taps, a significant amount of electricity is needed to run the circulating pump, and there will be heat lost continuously from the often poorly insulated hot water pipes spread around your home.

7.
Bye-bye, gas grid: cooking without gas

This chapter describes how to move on from that last gas appliance – the gas cooktop – and then how to disconnect your home from the gas grid altogether.

Fast times in the kitchen

For the past several decades, cooking on a gas-burning cooktop has been popular in Australia. However, these days we have another option that has numerous advantages over gas. These include it being:

* cleaner
* healthier
* safer
* cheaper to operate
* faster and more responsive.

It's called an electric-induction cooktop.

What is it? An induction cooktop uses electric magnets to turn your pot into the actual cooking element. You may need to see it to believe it, but when you turn on an induction cooktop the water nearly immediately begins to boil. When you are finished cooking and you shut off the cooktop and remove the pot, the glass surface will be warm to the touch, but you'd be unlikely to burn yourself.[1]

This is because the magnets work to heat the metal of the pot only and waste no energy trying to heat the glass surface.

A few years ago, we replaced our gas cooktop with induction. The other morning, I was using it to cook an egg with some spinach and tomatoes. It occurred to me that with the induction cooktop I had cooked the egg and vegetables faster than our toaster could toast my piece of bread. So next time I'll start the toaster before I start cooking the egg!

Is it really an induction cooktop ... or is it glass-ceramic?

An induction cooktop looks like a glass surface. It can also look identical to an electric-resistive or so-called glass-ceramic cooktop. Those older technology electric-resistive glass-ceramic cooktops will certainly cook your food with no need for gas. However, induction cooktops heat up and cool down more quickly, and are more controllable and responsive than glass-ceramic.

Electric-resistive cooktops became popular in Australia in the 1950s. Induction cooktops were marketed overseas as early as the 1970s but are only now becoming widely known across Australia.

When visiting clients in their home I'll often ask, 'Have you ever cooked with induction?' Initially the answer might be something like, 'Yes we did once, when we were on holiday.' However, after I ask a few questions such as, 'Did it take a long time to heat up?', 'Did it glow red hot?' or 'Could you have burned yourself on the glass surface?', it turns out that what they thought was an induction cooktop was a glass-ceramic cooktop. Be aware of the difference because they can look the same when they're turned off.

Can I still cook my favourite dishes on an induction cooktop?

Common questions arise when households contemplate banishing gas from the kitchen. 'Can I simmer rice?' 'Will I be able to use my wok?' 'Can I cook baba ghanoush or naan?'

More and more noted chefs and professional cooks, as well as more of our friends and neighbours, are letting us know that yes, you can still cook all those things.

In our house, I enjoy the digital controllability of our induction cooktop. I've come to learn that I'll make the coffee quickly on setting 'P', cook a steak on '12', steam some vegetables on '8', eggs on '6' and if I want a tiny amount of heat there's always '1'. An induction cooktop is so powerful that professional chefs warn that it's best to start with lower settings and work your way up as you get accustomed to how quickly the cooktop responds.

Get the gas out of your kitchen

The kitchen can be a risky and unhealthy place. And I'm not thinking only about my penchant for screwing up even the simplest dish.

Open gas flames in the kitchen are dangerous. Bleary-eyed in the morning, there was the time I set the droopy sleeve of my bathrobe on fire. As I age, I don't mind not having a flame in the kitchen anymore. I've never experienced a grease fire in the kitchen, but having no gas flame means the likelihood of a grease fire is greatly reduced.

My grandmother used to cook over coal in the kitchen. Back then, that was seen as a normal thing. When compared with cooking over coal or even wood, gas seemed to be an improvement. Gas was advertised as 'clean burning', but compared with what?

Burning stuff in and around our homes will negatively impact the air quality in and around our homes. When gas is used in a kitchen, some of the chemicals that can widely spread throughout your home include:

* water vapour/moisture/humidity
* carbon dioxide
* carbon monoxide
* unburned methane
* formaldehyde
* benzene
* tiny particulates (e.g. PM 2.5s, particles less than 2.5 microns in diameter)
* sulphur-containing mercaptans
* sulphur dioxide
* nitrous oxides ... and more.

Some of these contaminants can leak into your home continuously, even when your gas cooktop is shut off.

Beyond the water vapour and carbon dioxide, those other chemicals and fine particulates listed above can have a range of bad health impacts including asthma, other respiratory diseases, cardiovascular disease, leukemia or other cancers, neurological disorders and more.[2]

Having no gas flame means the likelihood of a grease fire is greatly reduced.

The health concern that seems to be of greatest interest to my clients is the way that gas cooking contributes to childhood asthma. The nitrous oxides produced when cooking with gas are the contaminant most strongly linked to asthma.[3]

Sometimes a client will express strong interest in doing things to their home such as improving insulation or getting off gas-fired heating. Then they'll say, 'I like my gas cooktop; we don't need to talk about that.' Initially I might not address this, but toward the end of a visit I'll ask, 'Does anyone in the household have any breathing issues, such as asthma?' Too often I hear the response, 'Yes, my child has asthma.' In this situation, ditching the gas cooktop quickly goes to the top of my client's priority list.

> Nitrous oxides are the contaminants produced by gas cooking that are most strongly linked to asthma.

Reducing health risks when cooking with gas

If the health impacts of gas cooking are news to you and you are still using gas, don't despair. Here are some ways to reduce respiratory and other health risks in the meantime, until you can get your gas cooktop permanently replaced.

A portable induction cooktop: the quick fix

Try a portable induction cooktop. These can be placed on a kitchen bench and plugged into an electrical wall socket, no electrician needed. The cost to buy a portable induction cooktop can be less than $100. Some councils, libraries and community organisations loan portable induction cooktops so you can give one a try.

We have seen MEEH Facebook group members arranging one, two and even three portable electric-induction cooktops in their kitchens until they can have a permanent induction cooktop installed, possibly as part of a kitchen renovation.

I once visited an Asian hot pot restaurant where every table had its own portable induction cooktop.

If you are using a portable electric-induction cooktop but then later do manage to install a permanent one, you could loan the portable to friends and other family members to try. Otherwise, keep it as a portable hot plate that can be used even outdoors. For some reason, we once decided to make soup on a 40-degree Celsius day. Rather than heating up the house, we made the soup outdoors using our portable induction cooktop.

Note that for your pots to work on an induction cooktop, they must be able to respond to a magnet. This means cast iron and many stainless-steel pots and pans will work, but aluminium and some stainless steel will not. Grab a magnet from your fridge door and test the base of your pots. If the magnet sticks, your pot should work on induction.

Turn on fans and open windows

When cooking with either a gas or electric cooktop, run your extraction fan vigorously (if you have one) and have the nearest window cracked open as well. Fans work better when they have access to a good supply of air coming in from an open window. Ensure that the extraction fan's grease filter is clean and not clogged up.

Ideally an extraction fan is vented to the outside of the home. However, some extraction fans merely recycle the effluent from cooking through a filter and then straight back out into the kitchen. Some kitchens have no extraction fan at all. Opening windows wide in this situation can help to purge cooking contaminants from the kitchen.

Run an air purifier

During bushfire-smoke catastrophes and the COVID-19 pandemic, many Australian households rapidly became familiar with portable home air purifiers equipped with activated charcoal filters and high efficiency particulate air (HEPA) fine particle filters. Especially in cases where a home does not have an externally vented rangehood extraction fan in the kitchen, an air purifier can be used to remove some of the contaminants generated by cooking. However, an air purifier may not remove nitrous oxides, the chemical produced by gas cooking that can lead to asthma.

An air purifier may come with a fine particulate (PM 2.5) monitor, which can alert you to sudden increases in this contaminant due to cooking or burning things in your home, or from your neighbours' home, or from uncontrolled bushfires and fuel-reduction burns beyond your suburb.

SEE ALSO *Draught-proofing and clearing the air (page 127) for more on ways we can keep our air clean*

Making the switch to electric cooking

Switching from a gas to a permanently installed electric cooktop involves spending money. You'll have to buy an electric cooktop, either induction or electric-resistive. You'll need to remove the gas cooktop with care, using a gas plumber to isolate the gas connection if the home is still connected to the gas mains. You'll need to have an electrician wire the cooktop on its own circuit back to your electrical switchboard. And you may need to alter your countertop if the cooktop you've selected won't fit where the gas cooktop used to be.

If it's going to take a long time for you to organise this, the portable induction cooktops that I described previously may provide a short-term solution.

There are limited programs underway in Australia where some governments and even some energy-distribution companies are helping selected households to transition away from gas cooking.

The Victorian government plans to offer rebates for households installing induction cooktops.[4]

Will you need three-phase power for the cooktop or anything else?

When considering installing an induction cooktop, common questions are: 'Will I need three-phase power? What is it anyway?'

Outside your home there are actually three power lines. The normal situation is that your home is connected to just one of these lines. That's called a single-phase connection. Most home owners will find that their single-phase connection can provide enough electricity (measured in amps) for an all-electric home, along with any electric car charging. In some cases, however, a home owner might need their existing single-phase supply upgraded to have a greater capacity (more amps). But this will still be a single-phase connection.

If for some reason you wanted or needed an electrical connection three times larger than normal, an electrician would connect you to all three of those power lines going by your house. That is known as a three-phase connection.

Is a three-phase connection necessary and what does it cost to get one? Taking the second question first, for a new-build home, I've heard of costs as low as $500 to connect to three-phase. But higher costs are also possible. For existing homes, I've not heard of any upgrades to three-phase being less than $2,000.

But is an upgrade necessary? How useful is three-phase?

Cooktop: There may be an induction cooktop out there that requires a three-phase connection, but I've not seen one. The thousands of homes I'm familiar with connect their induction cooktops with single-phase only.

Electric vehicle charging: Three-phase vehicle chargers exist, so if you would like to charge your electric vehicle really fast, consider upgrading. But most people will find a single-phase connection adequate for car charging.

Electricity export: Many solar PV systems have an export limit set to 5 kilowatts. That limit is set per phase. So if you have a three-phase connection, it is possible to export on each phase (each supply line), tripling the rate at which you can export electricity. In other words, your export limit would then be 15 kilowatts total.

Other big electricity loads are possible, such as arc welding and pottery kilns, but if those are your hobbies, you probably know more about three-phase power than I do.

My bottom line is that your home is unlikely to need a three-phase connection, but if someone wants to supply you one for not much money, feel free. For more on this topic, you can read an excellent article by Tristan Edis called 'If I go all electric, do I need to upgrade my grid connection?' at the *Australian Financial Review*.[5]

Induction cooking and pacemakers

People fitted with pacemakers need to keep their distance from the strong magnets used with induction cooktops. The British Heart Foundation recommends a distance of 60 centimetres.[6] If this distance cannot be maintained, the electric resistive (e.g. 'ceramic') type of cooktop may be preferred. Incidentally, induction cooktops are not the only magnet-containing devices that people with pacemakers need to be wary of. They might also need to be wary of hair dryers, electronic body fat scales, large stereo speakers and welding equipment.[7]

Your oven and other electric devices for the kitchen

Your all-electric home will have an electric oven. I've not encountered householders keen to hold onto their gas ovens, so I'll take that to mean I need not write more about these.

However, people do ask me what the most energy efficient type of electric oven is. Unfortunately, in Australia there is no energy star rating system for ovens. However, I can say that within a given brand of oven, the pyrolytic self-cleaning-capable ovens can be more efficient than a similar oven that lacks the self-cleaning capability. Whether you ever use the self-cleaning function will be up to you but, in any event, a pyrolytic oven can come with better seals, better insulation and better insulated glass than a similar, non-pyrolytic oven.

Our pyrolytic oven seems to be very efficient in that it's difficult to tell when it's on. The glass is not even warm to the touch. Our previous oven was less energy efficient and would readily heat up the entire kitchen. In those days we took care that children didn't bump up against the hot glass.

Should you have to use the oven in summer, a more efficient one will heat up your kitchen less and this may mean that you run your air conditioner less. That's a double efficiency benefit.

Some kitchens these days seem to be loaded with electrical cooking devices. So much so that the cooktop and oven might not even be the most used items. Air fryers, for example, are popular. Be careful with air fryers, however. From an air-quality point of view, air fryers can emit hazardous fine particulates (e.g. PM 2.5s). You may not see visible smoke coming from your air fryer, but an air-quality monitor would register a high figure for fine particulates. Ours does. So we now use the air fryer outdoors like a barbeque (you wouldn't barbeque indoors, after all), or we point the air fryer out the window or run it under the extraction fan with the fan on high.

It's good that we can clean up our air by getting rid of gas. However, we shouldn't then just introduce new health hazards with new electric cooking equipment.

Bye-bye gas grid, we're not coming back

In the homes I visit, the gas cooktop is often the last remaining gas appliance, with electric equipment already being used for home and water heating. Although cooking doesn't use much gas when compared with home and water heating, a household will still have to pay at least a dollar to the gas companies every day, until they disconnect from the gas grid.

Who wants to get a gas bill for the rest of their life? The prize of eliminating the gas bill entirely can provide the economic justification to take the final de-gasification step: replacing the gas cooktop with an electric option.

So perhaps congratulations are now in order, as your home no longer uses gas! You've electrified your living-space heating, water heating and cooking. What's next? There's one more thing to do: ring up your gas supplier and tell them you no longer want to receive a gas bill.

What happens after that is an evolving space in Australia.

Who wants to get a gas bill for the rest of their life?

A few years ago, any home permanently leaving the gas grid would have been viewed as unusual. However, these days the gas distribution companies are becoming more and more familiar with people leaving. It's clear now, with states and territories such as Victoria and the ACT declaring new homes won't be connected to gas, that the distribution companies must develop a plan to wind up the gas grid.

Exactly how, we (households, governments, regulators and gas distribution businesses) collectively do this and who will directly or indirectly pay for it isn't clear. So far, around Australia, households have been paying from $0 to $2,000 to disconnect from the gas grid.

The Victorian regulator announced in 2023 that individual households in that state should pay their gas distributor no more than $242 (including GST) to disconnect from the gas grid.[8] Elsewhere in cities such as Canberra or Sydney, electrifying homes have been charged up to $2,000. I don't understand why any household should be asked to pay anything for the final disconnection, after having done the right thing and gone through the effort of getting their home off a polluting and hazardous fuel.

Again, congratulations, you're off the gas grid. So, is that all there is to do?

Probably not! In the next chapters I describe how it can be worthwhile to improve what is known as your home's thermal envelope by reducing draughts, insulating, improving windows and window coverings, and more.

8.
Do you live in a leaky bucket?

In Chapter 5 (page 61), I wrote that the best way to *actively* heat and cool a home is often by using a reverse-cycle air conditioner. But I also mentioned that in our fairly mild Australian climates, if you take some basic steps you may be able to improve your home's 'thermal envelope' and *passively* keep your home comfortable on many days of the year without having to turn on any equipment.

What do I mean by the thermal envelope? This is the envelope you live in. Your home's thermal envelope separates your conditioned (heated or cooled) living spaces from the outside world. It includes the ceiling, walls and floor (which should have as few holes as possible), accompanying insulation, any building wraps and membranes, and then the outside protective materials of the home itself, such as bricks or weatherboards, tile or steel roofs, timber or concrete slab underfloor and floor coverings. Furthermore, it includes the window frames, window glazing and window coverings, inside and out. The thermal envelope may also be referred to as the 'building shell' or 'building fabric'.

Whether we call it the thermal envelope or the building fabric, it might be easiest to imagine these parts of your home as being like a leaky bucket.

In winter, as quickly as heat is brought into a home, it leaks right out again, just as water spills onto the ground from a leaky bucket. If your home needs draught-proofing, heat is going out

THE HOME WEATHER REPORT...

THE WARM AIR WILL MOVE OUT THROUGH THE WINDOWS, FOLLOWED BY A COLD FRONT MOVING IN THROUGH THE DOOR AND INTENSIFYING INTO A CATEGORY THREE ENERGY BILL...

with the draughts while cold air comes in. No home insulation is as perfect as a vacuum thermos bottle, so heat will continuously leak out through the ceiling, walls and floor. And the thin glass of windows can be the leakiest part of all.

The opposite thing happens in summer if we're cooling the home with air conditioning. Heat leaks in.

In Chapters 9 through 11, I describe ways to improve the individual components of your home's thermal envelope: draught-proofing, insulation, windows, and window coverings. What you should aim to do is to seal up your bucket as well as you can and then open windows or use some other means of ventilation to manage air quality.

SEE ALSO *Draught-proofing and clearing the air (page 127) for ways to manage air quality*

In the next section of this chapter, I describe computer-based tools used to assess a home's thermal envelope, as it is or as it is being designed to be, along with the heating and cooling equipment used within. These tools can furthermore be used to assess the economic and comfort value of improvements such as adding more insulation or window blinds.

Computer-based tools for rating home performance

Let's say you have just bought a home and you'd like to know, scientifically, if it will be comfortable and how much it will cost to heat and cool. In fact, wouldn't it be good to know those things about a home even *before* you bought it?

And now that you own the place, you'd like to make some changes. What will the impact of those changes be on energy bills and on your level of comfort? Is there some way to actually work this out without resorting to guessing?

Or let's say you (or your architect or designer) are designing a new home. How can you know if it will be comfortable to live in and what the energy bills will be like? How can design options be assessed, such as whether to make that north-facing window larger or smaller, or working out the value of adding eaves?

It turns out that there are computer-based tools that can answer those questions. However, these tools are not used as often nor as extensively as they should be.

NatHERS modelling: a powerful tool

Perhaps you've heard that now or in the not-too-distant future, new Australian homes (and major renovations) must achieve an energy rating of no less than 7 stars. This rating is determined using the Nationwide House Energy Rating Scheme (NatHERS) calculation method as part of the National Construction Code. The 7-star requirement for new-build homes supersedes the 6-star requirement introduced around 2010, which superseded the 5-star requirement of around 2003.[1]

Versions of the NatHERS rating tool are used for new-build designs across Australia using the information that is available on design drawings and documentation as input. Less often, NatHERS is also used for existing homes if their characteristics can be determined, such as the R-value of the wall insulation or the U-value of a window.

The full NatHERS scale ranges from 0 to 10 stars. Many older homes rate 2 stars or less and are often said to have the comfort of a glorified tent![2]

But what does this star rating business actually mean?

A new-build 7 star–rated home will need half of the energy for heating and cooling as will a 4.5 star–rated home, one-fifth of the energy that a 2-star home needs, and one-seventh of a 1-star home. These are huge differences and relate directly to the level of comfort you will experience in your home, as well as to the size of your energy bills.

What a well-performing home feels like

Some Australians who are building new homes, but who have never lived in anything other than an older, poorly performing home, may have no idea what it's going to be like to live in a 7-star home. The comfort! The lower heating and cooling bills!

If your home is not a good performer, and you've not had the experience of living in a well-performing home, I suggest you try to track one down. Find a holiday rental that claims to have a 7-star rating or better and book in for a cold winter or hot summer weekend. Such rentals do exist! We found a great one in The Little Green Cottage in Maldon, Victoria. Feel the difference, even without turning on the heater or air conditioner.

Given that many Australians live in climate zones that are relatively mild (when compared with other parts of the world), it's not that hard to achieve a 7-star home rating. It requires what I call normal insulation, double-glazed windows and not huge amounts of glass.

Limiting the amount of glass is the biggest stumbling block I see to new homes easily achieving 7 stars. We love our glass! SEE ALSO *Windows and window coverings (page 152) for more on windows*

Don't just tick the box, shoot for the stars

The NatHERS calculation tool is quite sophisticated. After the details of the home's thermal envelope are entered, a skilled modeller can tell you a pile of information, such as what temperature you can expect to feel in the middle of your lounge room at 10 a.m. on 22 January (under average weather conditions).

To improve accuracy in this regard, in 2022 the NatHERS climate files were updated to better reflect the recent past impacts of global heating. The CSIRO has furthermore created climate files to model what is to come as our Earth gets hotter and hotter.

However, too often a NatHERS assessment is used simply as a regulatory compliance tick-the-box exercise answering only a single question: does a new home design achieve the legal minimum 7 stars? It would be better if housing designers and owners/clients used NatHERS modelling to its full capacity to investigate design changes that might achieve better than the bare-legal-minimum requirement at little or zero extra building cost.

How to heat a home for $1.50 a day instead of $25

NatHERS will predict how much heating and cooling energy a new or existing home will require, in the units of megajoules of energy per square metre. But at this time, NatHERS can't yet take the next step and predict how much it will cost to heat and cool a home. Of course, the costs of heating and cooling depend not only on a home's thermal envelope and orientation (i.e. if it is north facing or south facing), but also on the equipment used to heat and cool it. However, a skilled NatHERS assessor should be able to convert for you the megajoules per square metre of annual heating and cooling costs, based on the equipment that is used to heat the house.

As an example, I mentioned before that some of my clients were spending $25 a day on gas-fired heating. Let's assume it's an average old and largeish home that rates only 2 NatHERS stars. If the home could be upgraded to the 7-star standard, the home would need one-fifth of the heating energy, slashing the gas bill from $25 a day to only $5 a day.

Then, as you saw in Chapter 5 (page 61), a home can be heated with reverse-cycle air conditioning at around one-third the cost of gas-fired heating. So $5 a day becomes only $1.66 a day.

When I visit clients who are spending $25 a day on gas-fired heating and I tell them I live in a 122-year-old weatherboard that has some upgrades but is far from perfect, and that we spend less than $1.50 a day for heating even on the rainiest and windiest Melbourne winter days, they find it unbelievable. But, as I have shown, it is possible.

The quality of the thermal envelope plays a large role, as does the choice of heating and cooling equipment.

Similar economic and comfort outcomes can be achieved with respect to cooling in summer or in hot climates year-round. A home with a poor thermal envelope, cooled with an old inefficient air conditioner, may run up electricity bills of perhaps $20 a day. But by improving the home's thermal envelope, using modern air-conditioning equipment and, if possible, adding solar PV panels on the roof, summer cooling bills can be slashed dramatically.

The Residential Efficiency Scorecard

At the home appliance or whitegoods shop, you've probably seen energy star ratings on refrigerators, air conditioners, clothes dryers and the like. In fact, you'll find them all in the database at the Australian government website: energyrating.gov.au.

However, for an existing Australian home anywhere (outside of the ACT), there is no mandatory home star-rating system that might alert a person about the good or bad performance of a home they are looking to rent or buy.

The ACT's Energy Efficiency Rating (EER) scheme is one example of mandatory disclosure that has been in effect since around 1999.[3] The United Kingdom and some other places around the world also have mandatory disclosure.[4]

In 2008, Australian federal, state and territory governments decided there should be such a mandatory scheme nationwide. To achieve this goal, a new home rating tool, named the Residential Efficiency Scorecard, has been built by the Victorian government, and has been tested and deployed nationwide, with more than 100 assessors qualified. However, no state or territory government has stepped up to require mandatory use of the Scorecard. Fifteen years on from the 2008 meeting of governments, we are still waiting for mandatory rating and reporting of home comfort and energy performance.

I am a qualified Scorecard assessor and have assessed more than 100 homes with this tool. So what is the Scorecard and how does it differ to NatHERS? Why do we need two rating schemes?

The Scorecard, like NatHERS, rates homes on a scale up to 10 stars. Like NatHERS, an average existing older home will rate around 2 or 3 stars and a pretty good home will be up around 7 stars. Scorecard is meant to use the same underlying CSIRO calculation methods as NatHERS.

Like NatHERS, Scorecard considers the home's thermal envelope. However, Scorecard then goes further to include details about the equipment that is actually physically in place for heating, cooling, water heating, lighting, whether there is a pool or spa, and whether the property has solar PV panels. Scorecard considers only the above-listed equipment because that remains with the home as tenants come and go or as a home is bought and sold: equipment that is bolted to the wall, so to speak. Equipment that is not permanently attached to the building, such as refrigerators or televisions, are ignored by Scorecard. Unlike NatHERS, Scorecard's stars don't relate directly to energy consumption; they indicate the actual costs to operate the home.

A 1-star Scorecard home may be a very large, poorly oriented home with an unimproved thermal envelope, may have no solar PV panels, and may incur energy costs of around $5,000 a year for living-space heating and cooling, water heating, lighting, and for operating a pool and/or spa.

A 9-star Scorecard home may cost nothing over the course of a year for heating and cooling, water heating and lighting, may be a smallish home with an excellent thermal envelope and well oriented to the sun, may have no pool or spa, and may have a large solar PV system to provide or offset electricity use.

A 10-star Scorecard home might actually operate cash positive each year (based on the equipment items that Scorecard takes into account), with very low energy purchases in total thanks in part to a large solar PV system.

A NatHERS assessment for a new home design can be done from an office. EER assessments for the ACT are also a desktop-rating exercise based on documentation provided. In comparison, a Scorecard assessment of an existing home requires the assessor to visit the property, sight and photograph a long list of the home's features, run the shower heads, look in the roof space to assess the insulation, see if the solar PV system is actually working, and more. During the home visit, the assessor meets with the householder and discusses in detail home comfort and energy

improvement options, based on the assessor's experience and/or the suggestions that the Scorecard computer tool comes up with.

Scorecard also rates, on a scale from 1 to 5, the home's ability to passively remain comfortable in winter and summer without using active heating or cooling.

If, after reading this book, you're still not sure what is the single next action you might take to improve your home, you might have a look at the Scorecard website and contact an assessor listed there: homescorecard.gov.au.

Why would we need thermal mass?

Many people will have heard architects and designers repeat something they heard about a home needing thermal mass. This theory claims that it would be a good idea, for example, to have a brick wall inside of a home to soak up heat from a heating system or a concrete slab floor to soak up the heat of the winter sun.

In my view, thermal mass is overrated.[5] Our own light and fairly tight Melbourne weatherboard home has little in the way of thermal mass. And yet we're comfortable and spend less than $150 a year for heating and cooling. Why would we need thermal mass? What a home needs is good insulation, draught-proofing, good window treatments, not too much glass, efficient electrical appliances, no gas, and solar PV panels if there is a suitable roof.

> In my view, thermal mass
> is overrated.

So where did the idea that thermal mass is good come from? My theory is that it dates to the centuries when people living in cold northern hemisphere climates would heat their homes with an open fire. Burning wood or coal in a fireplace gives off a lot of radiant heat. Thermal mass, such as stone or bricks nearby, could soak up this heat. With luck, as the fire went out overnight, the thermal mass would retain heat, meaning that the fire didn't need to be kept going all night long nor be urgently restarted in the morning.

Likewise in desert-like climates where the days are hot but the nights are cool, in the centuries before efficient refrigerative air conditioning and solar PV panels were available, a building with a lot of thermal mass, not unlike a cave, could provide a place of comfort.

A lot has changed since those days, but the idea that thermal mass is a good and necessary thing persists. Many of us have seen thermal mass backfire. Often my clients in older double-brick homes will tell me, 'Our home stays cool in summer for two or three days. But if the heat goes on for longer than that, the bricks heat up and it becomes quite uncomfortable.'

In your existing home, you won't be able to do much about adding or subtracting thermal mass. You have what you have.

But if you're planning a new home or major renovation, and your designer is a fan of thermal mass, ensure they carefully model its impact during both winter and summer. Compare that with the other design option, known as 'light and tight', which does not require thermal mass.

9.
Draught-proofing and clearing the air

Imagine it's a windy day. You're sitting at home with your windows closed. But you feel a breeze. Where's it coming from? Does your home leak air like a sieve or does it leak just a little? Most Australian homes I've visited are extremely draughty. This chapter describes how you can seal up the leaky holes, gaps and cracks around your home that allow draughts. And while you are draught-proofing, you can improve your home's air quality at the same time.

Draught-proofing brings big benefits

You may be aware that draught-proofing can be one of the best bang-for-buck ways to improve your home's comfort and reduce energy bills. But here's another benefit that many people hadn't thought about until we had those terrible bushfire and smoke events of 2019–2020: improving the cleanliness of the air you breathe. A draughty home can let air in that's polluted with smoke from a distant bushfire or from a neighbour's barbeque, as well as traffic pollution, seasonal pollen and other irritants.[1]

And there are two more reasons that living in a leaky home can have you breathing unhealthy air. Firstly, the air being drawn in can be contaminated as it passes through the holes, gaps and cracks from under the floor, through walls or from a roof space where dirt and mould have built up over time. Any air coming to you through an old wall vent or past a leaky downlight in the ceiling can't be guaranteed to be 'fresh' uncontaminated air.

Secondly, a leaky home can allow moisture (water vapour) generated by the home's occupants to move into walls, where it may promote mould growth in the hidden spaces of the wall structure, behind plaster and in insulating materials.[2] The bad news is that through the same gaps and cracks, mould spores can then move back into your home, leading to poor health.

Why so leaky? Feeding the fires

In bygone days, Australian homes were designed and built to be very draughty with large volumes of air leaking in and out of homes, especially on windy days. Why? So that all winter, wood or coal could be burned in fireplaces, or gas burned in gas-fired heaters or gaslights. Fires need air to burn. Also, massive air leakage helped to carry away harmful combustion contaminants, such as smoke or carbon monoxide. Eventually, uncontrolled air leakage became part of the Australian building standard on the thinking that it might reduce the number of people killed by poisonous carbon monoxide coming out of a gas heater.

In the past, building a leaky home was also a crude but available way to reduce the build-up of moisture and other contaminants in a home, with an aim of limiting the health impacts caused by cigarette smoking, moulds, dust mites, bacteria and viruses.

Even though having a draughty home isn't a regulatory requirement today, many modern homes continue to be draughty due to poor construction practices, shortcuts, oversights and adherence to out-of-date practices.

Avoiding death by carbon monoxide, and other risks

Before draught-proofing your house, eliminate sources of carbon monoxide. Also ensure the risks from other air contaminants, such as moisture, are managed. Read this chapter to understand how to manage these risks.

Quantifying air leakage with the blower-door test

The draughtiness, or you could say the leakiness of a home, can be scientifically tested with a blower-door test.

 There are lots of good videos demonstrating the blower-door test on YouTube.

With the blower-door test, a technician opens your front door and fixes a blower there. They then seal off the rest of the door space. All other external doors and windows are closed during the test. They then turn the blower on, which gently tries to suck all the air out of your home. Because no home is perfectly airtight, air rushes in around every window, under the laundry room door, from above the big gap over the fridge, through wall vents, down chimneys and backwards through bathroom extraction fans. The technicians will use a smoke machine to provide visible proof of the air rushing in. You might feel this air with your fingertips or the back of your hand. A thermal imaging camera is sometimes used to obtain a visual image of cold or hot air rushing into a home.

The blower-door test simulates a very windy day, which is when your home is under pressure and will leak the most.

Using pressure measurement devices and the dimensions of your home, the blower-door test provides official figures and results of your home's leakiness. This includes an estimate of how many square centimetres (or square metres!) of holes, gaps, cracks and other openings penetrate your home, as well as an estimate of how many air changes per hour (ACH) your home will experience on a windy day.

Where to look for the holes, gaps and cracks

An older Australian home may have 1 square metre or more of holes, gaps, cracks and openings, exposing your living spaces to the weather at all times. This is easy to imagine if you mentally sum up the following possible leaky places:

* coal/wood fire chimneys (some old homes may have as many as four old fireplaces and chimneys open to the sky)
* older gas heater flues

- air leakage via the evaporative cooling system
- bathroom, toilet and kitchen extraction fans with no back-draught dampers
- old-style wall and ceiling vents (intentional holes formed through the plaster)
- gaps around window and door architraves that a builder or painter neglected to caulk up
- windows in a toilet or laundry that may feature a screen instead of glass
- gaps where the windows and doors are meant to seal
- unnecessary ventilation gaps around downlights (especially the older halogen type)
- the big hole behind and over the refrigerator or behind and over a clothes dryer
- gaps along skirting boards and kitchen cabinet kickboards
- gaps down to the ground where tongue-and-groove floorboards have broken
- holes and gaps associated with failed heating/cooling air ducting
- gaps or massive holes around ducted heating/cooling inlets and outlets
- gaps around plumbing pipes where they go through the wall or floor
- cracks in plaster, and more!

With the blower-door test, a very leaky home might score 60 ACH, meaning that on a windy day all the air in your home is being swept away every minute, along with all the money you paid to heat or cool that air. There is no way to be comfortable and have low energy bills in a home like that.

A modern home built well, or a draught-proofed older home, may be ten times better than a leaky house, scoring as low as 6 ACH. The European Passivhaus standard aims to do ten times better again and to score less than 0.6 ACH.

But you don't need to have a blower-door test done, nor do you need to build your Australian home to the European Passivhaus standard to be comfortable and have low energy bills. An Australian home or renovation that has been properly draught-proofed and constructed to at least the new 7-star minimum legal requirement will provide you with a home that is comfortable in all seasons while having low energy bills.

For existing homes, what does need to happen is that holes, gaps and cracks are identified and sealed up. This can be a DIY exercise. I list below where you can find videos and other information online to help you. Or you can hire a handy person or ideally a professional draught-proofer to identify the gaps in your home, provide an itemised quote and to seal up those gaps.

Draught-proofing tips, information and resources

Renew Magazine, published by the not-for-profit organisation Renew (formerly known as the Alternative Technology Association or ATA) features many buying guides. These include a draught-proofing buyer's guide. Many of *Renew*'s buying guides are available on the internet.[3] Hard copies of *Renew Magazine* can be found in libraries.

The founders of the draught-proofing company ecoMaster, Lyn and Maurice Beinat, have for many years been uploading detailed 'how to draught proof' videos to their website.[4] The website Green-It-Yourself (greenityourself.com.au) explains draught-proofing techniques and stars ABC presenter and former 'Carbon Cop' Lish Fejer.

Next, I describe how to keep the air in your home clean and healthy.

Keeping the air in your home clean

More congratulations! You've draught-proofed your home and now it's more comfortable, cheaper to run and you've reduced the rate at which air contaminants, such as smoke, can come inside.

But you're not quite done. Inevitably more air contaminants will continuously build up within your home: carbon dioxide and moisture (water vapour) as a minimum, and also possibly fine particulates (smoke) from cooking and from burning things such as candles. There are also other contaminants including volatile organic compounds (VOCs) from plastics and other materials, mould spores, dust and dirt particles and, in some parts of the world, even radioactive contaminants.[5]

In the past, you may have unknowingly been relying on large volumes of air leaking into and out of your home to purge these contaminants. However, in less leaky homes we must now spend some effort to:

* minimise the presence of these contaminants
* purge them out of our homes by ventilating (e.g. opening windows and using extraction fans)
* use contaminant removal equipment, such as dehumidifiers and air purifiers, in some cases.

In the next section, I describe general ventilation strategies and then I individually address what can be done about the contaminants listed above.

Moving from a regime of air leakage to ventilation puts you in control

As we and our pets breathe, we expel moisture and carbon dioxide. In a draught-proofed home, we need to purge out these contaminants as a minimum. In most homes, this will mean judiciously opening windows and also running bathroom and kitchen extraction fans.

What do I mean by 'judiciously'? Open windows when you need to. But how will you know when you need to open windows? This is a very good question, and I present some ideas later when I discuss the individual contaminants. But as you will see, it can be challenging keeping moisture and carbon dioxide down at the best low levels, so it means windows need to be opened a lot. Not during the worst weather conditions but very often at other times.

It's at this point people often ask: 'Why did I go to the trouble of draught-proofing my home only to then open the windows much of the time?' The point is that a leaky home leaks all the time, 24/7. This is known as 'uncontrolled air leakage'. After you have draught-proofed your home, you can move into the regime of 'managed ventilation'. This means you will be in control of your home's ventilation. Open windows when you want; close them tight during the worst weather or when there is smoke or other pollutants outside.

You need not worry too much that opening windows will cause your heating and cooling bills to climb too high. By following the tips in this handbook (especially those about draught-proofing, insulating, window coverings and heating cheaply with air conditioners), your heating and cooling bills should be under control.

Among Australian housing professionals, discussions are beginning about whether continuous ventilation should be suggested or required in our homes.[6]

To ventilate continuously in the most thorough and automatic way, some modern homes are equipped with mechanical heat recovery ventilation (MHRV) equipment. This equipment can include various levels of air filtering and purification. Nevertheless, the equipment may need to be shut down during times of high outdoor humidity or smoke pollution if it cannot adequately condition the incoming air. MHRV equipment is available either in a ducted form, which is most readily applied to new construction, or in paired decentralised systems that require no ductwork and can be retrofitted to existing homes.

 Two companies I recommend for sourcing MHRV equipment are Zehnder (international.zehnder-systems.com) and Laros (laros.com.au).

Managing moisture (humidity)

Surprisingly to some, since Australia is often characterised as a 'desert continent', the moisture/humidity level in many Australian homes is often too high. In some homes the humidity is too high in winter; in others it's too high in summer, and for some homes it's both.

The recommended relative humidity (RH) range is between 40 and 60 per cent, with 50 per cent RH being the sweet spot.[7]

Unfortunately, I often see homes (including my own) where the relative humidity can reach 80 per cent or more. Levels above 50 per cent can support the growth of dust mites, bacteria, viruses and mould, which then leads to a number of health issues including asthma.[8]

The presence of excessive moisture (humidity) in your home can be checked with a $20 humidity sensor. In winter an obvious telltale of high humidity levels is when moisture condenses on the inside surfaces of cold window glass or even on cold plaster ceilings or walls. If a surface is cold enough and the humidity in the air high enough, dew point conditions will be reached and that's when water droplets form on those surfaces.

 A good, affordable humidity sensor is the Thermpro Digital Hygrometer Indoor Thermometer, which you will find on Amazon.

Where does all this moisture come from? Essentially from our breath, as well as from the breath of any animals that live indoors with us. We also add moisture to the air when we cook, bathe or shower. Any water that is added to houseplants or fish tanks ultimately ends up back in the air. Hanging wet laundry around the home with the windows closed dumps a large amount of moisture into your air, as will a non-condensing 'vented' clothes dryer that isn't vented to the outdoors. Moisture can enter a home structure from roof or gutter leaks, and especially if proper drainage around the building is not maintained.

So, what are some tips for winning the war against moisture?

Be a fan: Run bathroom extraction fans and cooktop rangehood fans when showering or cooking to purge out moisture and other contaminants. Ideally, these fans should be ducted to the outside of the home. Fans work best when a nearby window is also opened, at least slightly. Clean the grease filters in the rangehood regularly as well as the dusty intake grilles of bathroom fans.

Use a Showerdome: Over any enclosed shower stall, consider fitting the commercial product Showerdome. It's essentially a lid over the shower, raising the question in my mind of why don't all shower stalls come with lids in the first place? Rather, in modern homes built to impress, we find massive open showering rooms. Showering in a space that is enclosed on all sides means that when you step into a shower it's already warm, so you don't use as much hot water to heat yourself and the entire bathroom up. All moisture is confined to the shower stall itself, meaning there's less chance that mould will be formed anywhere else in the bathroom. The bathroom mirror doesn't get steamed up and a Showerdome minimises the need to crack open a window and run the extraction fan.

Some people have used a big piece of plastic corflute, such as a leftover political campaign poster, to test if they'll like the Showerdome effect. However, the dome shape of a true Showerdome will give the best result. After this test, most people find they do like it.

One of our daughters went to work in London for two years. When she returned, we found out she hadn't missed Mum, Dad or the dog. She missed our Showerdome.

 If you are interested in looking at a Showerdome for your home, go to showerdome.com.au.

Don't dry clothes indoors: And here's a big one. You might say I am calling for an Aussie icon to be put to death. Avoid hanging laundry around your home to dry while you have your windows closed. This can lead to excessive moisture being added to the air.[9] Dry clothes outdoors when possible.

When drying outdoors isn't possible, consider using an efficient 'heat pump-condensing' clothes dryer. These dryers condense the moisture from your sheets, towels and clothing into a bin from which the collected water can be poured down the drain. With an energy rating of 7 to 10 stars, heat pump-condensing dryers may use as little as one-quarter of the electricity that was needed to run an old-style 1-star 'vented' dryer and therefore won't have a big impact on your electricity bill.

If you don't have a heat pump-condensing clothes dryer, another method by which moisture from clothes can be collected and drained is to hang clothing in a small room while running a dehumidifier.

Certainly, hang clothes outdoors to dry if and when possible. Take advantage of the free solar and wind energy out there just waiting to be used!

Wipe down wet windows/dehumidify: If moisture condenses on cold windows, walls or ceilings, wipe it up and put that moisture down the drain. In this way, cold windows are actually acting as see-through dehumidifiers. In homes where condensation on windows is very common and excessive, some people quickly suck up the moisture using electric window vacuums, such as those used by professional window cleaners.

Find examples of electric window vacuums at kaercher.com/au/home-garden/window-vac.html.

Should summer or warm weather humidity levels be higher than the healthiest range, your air conditioner can dehumidify the air. Setting your air conditioner on COOL or DRY mode in humid weather can condense water out of the air. This water then dribbles out of the air conditioner's condensate drainpipe outside of your home.

However, in winter the COOL or DRY settings might not be effective at removing moisture or may chill a home beyond what is comfortable. In this case a desiccant-type dehumidifier can be used in cooler weather conditions. This type of dehumidifier uses a rotating wheel of desiccant material (think of the product 'DampRid') to absorb water from the air while also adding some heat to the room.

Run the air con on HEAT mode: One reason that humidity levels can get too high in bedrooms in winter is because we tend to underheat our bedrooms. With an efficient reverse cycle air conditioner on HEAT mode, don't be afraid to warm up a bedroom now and then, even with the window open, if that is the only way you can move moisture out of the bedroom and keep humidity levels near the healthy range.

Managing carbon dioxide

As discussed, humans and pets continuously breathe out carbon dioxide. Clearly, there's not much to be done about these sources of carbon dioxide in your home other than encouraging people and pets to spend more time out of the home.

The amount of carbon dioxide in our Earth's atmosphere has risen from a concentration of 280 parts per million (ppm) in pre-industrial times to around 425 ppm today. It will keep rising as humans continue to burn fossil fuels and plant matter and as dangerous 'feedback loops' kick in with further climate disruption. If you spend time in a relatively small and airtight room, such as a draught-proofed bedroom with the windows and doors closed, you may find the carbon dioxide level can quickly rise to 1000 ppm and beyond. The health impacts of elevated carbon dioxide continue to be studied.[10] At 1000 ppm and beyond you may have feelings of restlessness or drowsiness and experience headaches or poor concentration.[11] You might not get a good night's sleep.

 I bought a quality Aranet brand carbon dioxide monitor for around $300: aranet.com/products/aranet4. (Be aware that there are far cheaper devices advertised on the internet, but they might not even measure carbon dioxide at all.) Having this device has led to me opening the windows in our home more often than I used to, to lower carbon dioxide levels. Ensuring you purge out carbon dioxide can mean you are also giving other contaminants an opportunity to leave your home.

Beyond sensible opening and closing of windows, what else can be done to manage carbon dioxide build-up? For most existing homes, probably not a lot. Running extraction fans in the bathrooms, toilets or kitchens is another way to move more air out of the home. It's not like the movie *Apollo 13*, where you can ask NASA to quickly rig up a carbon dioxide removal system.

A German practice of routinely allowing a large flush of air through the home is known as *stoßlüften* or shock ventilation.[12] With this method, air contaminants may be removed without a huge or continuous loss of heated or cooled air.

Managing fine particulates and other pollutants from burning stuff

When we cook (even in all-electric homes) and when we burn materials and fuels indoors, such as incense, candles, gas or wood, we contaminate the air in our homes with a range of pollutants, from asthma-inducing nitrous oxides to fine particulates (e.g. PM 2.5 or particles smaller than 2.5 microns in diameter) that are linked to heart and lung disease.[13]

SEE ALSO *Get the gas out of your kitchen (page 105)*

 Our Phillips air purifier displays the level of PM 2.5 particles in our home's air. I also use a portable air quality monitor (the IKEA 'Vindstyrka' bought for around $60) indoors and outdoors. It provides information on PM 2.5 particulates, relative humidity and temperature, and confirms what my nose tells me when my neighbour is burning wood or there is other burning going on in and around Melbourne.

To remove contaminants when cooking, make sure rangehood grease filters are clean, run the extraction fan hard, and have a window or door cracked open nearby to allow air to easily flow to the fan. Avoid burning, smoking or vaping anything in or around the home, including fuels such as gas, LPG, wood, ethanol, kerosene, charcoal, hydrogen or coal, candles, tobacco or incense.

If you are not able to get your home off gas straight away, remember to have any gas heaters checked no less frequently than every two years to see if they are producing poisonous carbon monoxide. If carbon monoxide poisoning is a risk, you may wish to invest in a carbon monoxide alarm.

One of my clients showed me the carbon monoxide alarm they had bought after he and his wife found they had been falling asleep on their couch at 6.30 in the evening. They felt they were too young to be going to sleep that early. It turned out their gas heater was trying to kill them.

Other air contaminants

We also bring air contaminants into our homes with things such as food, cleaning chemicals, oils, paints, hobby items, building materials, fabrics, and furniture. Housemates and co-occupants can spread bacteria and viruses. Opening windows and doors allows pollen, dust, smoke and moisture in. Mould spores from moisture trapped within and on walls can develop over time if not addressed. Animals such as insects, spiders and pets bring in more air contaminants.

Avoid using cleaning, hobby or other materials in your home that can contaminate the air. Choose paints and other materials that will have the least air contamination impact.

Removing contaminants with air filters and purifiers

Heating and cooling equipment (e.g. air conditioners, ducted heating and cooling systems) should be equipped with an air filter capable of removing larger particles, not only to somewhat clean up the air you breathe, but also to keep ducts, blowers and fans, heat exchangers and other parts of the system clean. Home ventilation systems, such as the MHRV equipment discussed previously, will include some level of air filtration.

To further clean your air, you can use an air purifier, particularly one with the highest level of filtration, such as a high efficiency particulate air (HEPA) filter, which can remove very fine particles. Air purifiers may come equipped with a sensor and read-out that can give you an idea of the level of dangerous fine particulates that may be present in your home. Air purifiers became scarce items in Australia in 2019–2021 as we dealt with bushfire smoke and the pandemic. Obtaining one before the next air quality crisis might be prudent. Ventilation systems or air purifiers may also include carbon filters that absorb some chemical contaminants, such as volatile organic compounds (VOCs).

10.
Insulation: a priority

Insulation works. It is one of the best things you can do to help improve your home's thermal envelope and make your home more comfortable and affordable. Your home should have insulation applied to the roof space, the walls and the floor. In some places in Australia, such as Victoria, for more than a decade it has been a minimum legal requirement to have insulation applied to each of those parts of a home.[1]

Unfortunately, I've been in many Australian homes where there is no underfloor insulation, no wall insulation, and the roof space insulation is thin, patchy, incomplete or disrupted.

If the insulation of your home is incomplete, I recommend you get some insulation installers around to quote on improving this. Or sometimes, you can DIY.

Perfect insulation in your roof space

The roof space of every Australian home should have a thick and perfectly continuous layer of insulation. This is likely to consist either of fibreglass or polyester insulation batts, or a blown-in insulating material.

The best performing roof space insulation should be fluffy and not crushed or compacted. Why? Because it's not actually the fibreglass or polyester that is doing most of the insulating, it's the air trapped within the structure of the fibreglass or polyester that is the important insulator. This is the same as with a down puffer jacket. It's not the downy feathers that insulate so much, rather it's the air that the feathers trap that keeps you warm.

Still air, trapped in place, is a very good insulator. We'll see this again when we discuss double-glazed windows. When that still air (or even better, the gas argon), is trapped between two panes of glass, it insulates very effectively.

The cost-to-benefit ratio for improving your roof space insulation can be quite good, meaning you will be more comfortable, and save money on your heating and cooling bills, by making your roof space insulation thick and perfectly continuous. Make it so!

I've climbed up my ladder and stuck my head up into hundreds of roof spaces. Rarely do I see perfect insulation. And near perfection is what's needed. If the insulation in a roof space covers, say, only 95 per cent of the ceiling, the 5 per cent of gaps wrecks performance massively.

The analogy I use is that you might have a thick doona on your bed, but if it doesn't cover your leg, you're not going to have a comfortable night's sleep.

I can readily see the impact of gaps in roof space insulation using a thermal-imaging camera. In summer under the vicious Australian sun, your home's roof space is blazingly hot. I know, I've been up there. Back down in the lounge room, when I point the thermal-imaging camera at the ceiling, any gap shows up as a bright red spot. It's almost like someone has mounted a radiant heater on your ceiling. Often the only quick-fix defence against these phantom radiant heaters on your ceiling is to activate an air conditioner. A more cost-effective idea is to fix the insulation.

The halogen downlight scourge

Why are there so many gaps in our roof space insulation? A key culprit was the trend in the 1990s and afterwards to have halogen downlights spread across our ceilings like the Milky Way. These highly inefficient lights, which were essentially heaters that happened to give off a bit of light and ran hot enough to cook a chicken, have caused many fires.[2] To reduce fire risk, insulation had to be kept well away from halogen downlights.

To add to the damaging impact of halogen downlights, the light fitting often came with an air gap to allow air to ventilate the light and keep it cooler. This air gap enhanced a home's air leakage and furthermore allowed roof space filth to drop down into your living areas.

The owners of any home that still has halogen downlights or even some light-emitting diode (LED) replacement downlights that also leak air, should consider if they need and want to have downlights at all. Can some or all of your downlights be eliminated, and the ceiling holes fixed by a plasterer? Where you wish to have downlights, they should be upgraded to efficient, airtight and safer LED lights. Safety-rated LED downlights can allow insulation to be placed right up against and even over the light fitting (check the documentation that comes with the light).

After lights have been upgraded, the roof space insulation can be improved.

Can some or all of your downlights be eliminated, and the ceiling holes fixed by a plasterer?

Even though I can't imagine that obsolete halogen downlights are being installed in any new homes today, some builders are still carrying on with the practice of keeping insulation well away from any and all downlights, even away from the safety-rated LEDs. So sadly, even a brand-new home may need to have its roof space insulation immediately improved.

Avoid chopping holes in your ceiling for ducts and ventilation

Seriously, don't cut holes in your ceilings! Beyond the downlights, a second way for penetrations through the plaster to occur is for duct outlets and inlets used for heating and cooling. Clearly, where there are ducts coming through the plaster, there is no insulation and there is a good opportunity for more air leakage. Where heating/cooling systems that involved ducts in the roof space have been eliminated, the holes in the ceiling can be plastered up and the insulation improved.

In decades past, prior to the use of extraction fans, ventilation holes were cut into kitchen, bathroom and toilet ceilings. Since in most cases the home will now be fitted with an extraction fan, these ceiling vents can be removed, the plaster repaired, and insulation installed where the hole used to be.

DOWNLIGHTS   UP LEAKS

Fans and 'whirlybirds' won't keep you cool

Often people are aware that their roof space gets very hot under the Australian summer sun. A neighbour's roof may feature a 'whirlybird', spinning around, presumably removing some heat from the roof space. Which leads people to then think, 'Maybe I should also get a whirlybird to cool my roof space.' Don't bother. A whirlybird is a moving piece of equipment that could fail or blow away someday; it's a hassle you don't need.[3]

A thick and continuous layer of insulation in your roof space is what you need to protect your home on the hottest days. If you were trying to cool your roof space by moving air around, engineering calculations show that it would require not a whirlybird but rather something resembling a jet engine to do an effective job.

> A thick and continuous layer of insulation in your roof space is what you need to protect your home.

Roof spaces may need to be ventilated – not to remove heat but rather if there is a concern about moisture condensation. I'll say that this building integrity topic is beyond the scope of this book, but one tip is to see if the source of the moisture entering a roof space can be eliminated.[4]

Roof membranes

Roof membranes, often referred to as sarking or wrap, may be found beneath roof tiles or steel. They may contain aluminium and, if so, will prevent a lot of the sun's heat from reaching your roof space. However, you don't live in your roof space so, again, the best and most effective defence against heat entering through your ceiling in summer or heat leaving your ceiling in winter, is a thick and continuous layer of insulation.

Roof membranes can have benefits beyond improving thermal performance, such as managing rainwater and water vapour. However, again, those building integrity topics go beyond what I'm covering in this book.

Don't have a roof space?

Many homes or sections of homes have no accessible roof space, as they have a flat or skillion roof instead. I've often been able to identify insulation flaws in these roof types using a thermal-imaging camera under appropriate weather conditions. So how can these be insulated or have their insulation improved if needed?

If the insulation in such a roof is less than it could be, there are some insulation companies in Australia that can blow in a fluffy yet solid insulating material.

Other homes may have a raked or so-called cathedral ceiling, perhaps with exposed beams. Some owners of these have opted to install insulation and plasterboard inside the home, sacrificing some or all of the exposed beam effect.[5]

Wall insulation

Ideally, Australian homes should have a very good layer of insulation in the walls, but many will not.

State and national construction codes that were in force when your home was built may indicate what level of insulation could be in the walls. Removing plaster or external siding during a renovation or repair project will reveal how much insulation is actually in there. Peering behind an electrical receptacle cover, if safe to do so, is another way to look for wall insulation. If the weather conditions are right, thermal imaging can be used to determine if there is insulation in walls and how consistently it has been applied. Are there missing sections or gaps?

For homes that have no wall insulation, it can be a major renovation exercise to remove plaster or weatherboards, and then to install a wall membrane and insulation batts. As mentioned previously, there are some insulating companies in Australia that can blow a fluffy but solid insulating material into walls either by drilling holes outside of the home at the mortar joint of a brick wall, through weatherboards, or through the plasterboard inside the home.

At our old Melbourne weatherboard house, we managed to insulate all but one of our old walls either by stripping off the lathe and plaster inside, or by removing some rotting weatherboards on the outside.

We were left with one wall that I eventually confirmed was not insulated. One reason I suspected it was not insulated was because in winter it felt icy when compared with all the other walls. We had insulation blown into that wall and now the temperature of that wall in winter is no different than any other wall. It was an important wall to insulate because it's in the room which has our main reverse-cycle air conditioner. Since the time we insulated that wall, that whole part of our conditioner doesn't have to work hard to achieve a comfortable room temperature.

Underfloor insulation

A major thermal and comfort flaw of our home is its lack of underfloor insulation. Our house was built on stumps and is close to the sand, so I am still digging! I'm always a bit jealous when I visit some clients' homes that have plenty of underfloor clearance. Retrofitting underfloor insulation in these homes would be so easy.

Any Australian homes that are likely to require extensive heating and/or cooling should have underfloor insulation. In some states and territories, installing underfloor insulation in new homes has been required by regulations for more than a decade.

Homes with underfloor clearance of at least 600 millimetres should be able to retrofit insulation. The preferred material and method is to staple long lengths of polyester insulation to the floor joists. This is the method that has been found to resist animals and gravity. Avoid using fibreglass, polystyrene or aluminium foil underfloor, unless extreme diligence is applied to ensure it stays in the required position, tight against the floor.

 Refer to information and videos about polyester underfloor insulation at the ecoMaster website: ecomaster.com.au. They are the gurus of this insulation method.

Unfortunately, for existing homes on concrete slabs, retrofitting underfloor insulation isn't possible.

A member of the Facebook group MEEH was partway through insulating their floor during winter. On the insulated part of their floor, they measured a temperature of 16 degrees Celsius. On an uninsulated part of their floor, they measured 12 degrees Celsius. Insulation works!

11.
Windows and window coverings

Let's assume you're heating your home with efficient electrical appliances. You've draught-proofed and insulated your home. So now, thinking about what remains to be done to improve your home's thermal envelope, that leaves the windows.

We love our windows. They're often very big and sometimes even HUGE. In a client's home, my thoughts might turn to wall insulation. But on looking around, I realise some Australian homes hardly have any walls at all. Everywhere it's glass.

A single pane of glass provides little protection from the heat and cold outside. Many Australian homes are uncomfortable

and have big heating and cooling bills mainly because of their large and poorly performing windows.

However, because it can be expensive to upgrade windows, let's first discuss what can be done with coverings for the windows inside and out, such as drapes, blinds, eaves, awnings and even bubble wrap. Investing in very good window coverings can often be more cost effective than upgrading the glazing. This can often be a 'no regrets' approach because, even if you decide later to upgrade some windows, you'll continue to want to use your good window coverings anyway.

Warm and cosy coverings for the inside of your windows

Installing excellent window coverings inside your home, and then actually using them, will help to keep a home warmer in winter and cooler in summer.[1]

If you are designing a new home, or a new room, or just thinking about revamping your windows, always think about what sort of window coverings you will get inside and outside, and how they will be attached. Will they be attached to the wall, to the outside of the window frame or to the inside of the window frame? How will this be done? Can this be done in a way that doesn't block off the light coming through the window, and in a way that allows the window to be easily opened and closed?

In order of better-to-poorer winter and summer thermal performance, the interior window covering list goes like this:

1. thermally lined heavy curtains/drapes complete with pelmets
2. honeycomb/cellular/accordion blinds
3. roman blinds

4. plantation shutters
5. roller/holland blinds
6. vertical, venetian or slat blinds.

To think about which window coverings are the best thermally, imagine you had to sleep outdoors on a cold night. Which of the above would you prefer to use as a make-shift blanket? Probably not aluminium venetian blinds!

Curtains/drapes and pelmets

For energy saving and comfort, the best performing interior window coverings are heavy curtains/drapes complete with pelmets. Drapes should completely hang to the ground and wrap around the sides of windows when drawn closed.

A pelmet is a structure or arrangement that boxes in the top of the curtains/drapes. It's not just for style, nor a throwback to the eighteenth century. People in the eighteenth century actually understood how valuable a pelmet can be. A pelmet helps to keep the air in your room from reaching cold window glass in winter. Without pelmets, heated indoor air rises and travels across the top of the drapes to find the cold windows, which then chills the air and makes it heavier. Cold heavy air falls to the floor and, in this way, you get a circulating internal air current that can make it feel like a home still has draughts even after it has been draught-proofed. Pelmets with drapes wrapping around the edges of the windows and hanging to the floor ensure this air circulation phenomena never gets started.

When you wish to open the drapes and let the full light in, it's ideal if there is an extended curtain rod and a space off to one side or the other of the window (or both) where the drapes can be 'parked'. This means you'll retain the full view out the window and not block any of the light nor any of the sun's heat coming in when you want it to come in.

However, not everyone likes the look of drapes, and they can be expensive to buy and install. Next on the performance list are honeycomb blinds, so let's talk about those.

Honeycomb blinds

Honeycomb blinds might also be referred to as cellular, accordion or pleated blinds. These blinds are arranged like an accordion, with two or more layers of fabric, that when opened trap air within their cellular structure. Trapped air is an excellent insulator.

Honeycomb blinds can be translucent – allowing light through – or made of a material that totally blocks out light for sleeping, if that is what is needed. Some honeycomb block-out type blinds may have an aluminium coating within, further stopping heat transfer and improving thermal performance.

Attention should be paid to how honeycomb and all types of blinds will fit, such as either close to the window frame or within the window frame and close to the glass. Achieving a pelmet-like arrangement is the best outcome, as the blind is able to trap a still layer of air next to the glass with minimal gaps all around.

If you got it, don't forget to use it

Inside our house, we have three types of permanently affixed window coverings: holland/roller blinds, roman blinds and honeycomb blinds. These cover some double-glazed and some single-glazed windows. But then it's important that I go to great lengths to actually use these blinds. The story is that I'm able to avoid a gym membership because I run around our house twice a day opening or closing 28 window blinds. Such effort is, in fact, critical to achieving greater comfort and low heating and cooling bills. If opening and closing blinds feels like a chore, know that more and more homes use electrically powered mechanisms to open and close blinds with the push of a button.

I don't often sit with clients in their homes after dark. However, once I was with a client as late afternoon became a winter evening. The client complained of having high heating bills and being uncomfortable. I said, 'You have some nice blinds here, let's pull them down.' The client responded, 'Ooh, I never do that.' So, we pulled the blinds down and about ten minutes later the client said, 'Yes, it does feel more comfortable here than it normally does.' On a cold night, blinds stop your body heat from radiating right out through the window to outer space. Blinds work!

Low-cost ways to keep heat in

Sticking bubble wrap packing material to your window glass is another way to prevent heat from going out through a window. Renters and budget-constrained households might not be able to go out straight away and install honeycomb blinds. A cheap fix can be bubble wrap. You might have already collected a supply of free bubble wrap packing material, otherwise it's not expensive to buy. Bubble wrap stuck to windows creates an effect similar to double-glazed windows. A layer of still air is trapped next to the glass or in the plastic bubbles themselves, air being an excellent insulator. Bubble wrap will also allow some light through. So this can be a good solution, for example, in a child's bedroom where there might not be an attractive view through the window anyway.

In her rented unit, my daughter and I recently bubble wrapped my grandson's bedroom window. Bubble wrapping made the room warmer. It also reduced the moisture that was condensing on the single-glazed aluminium-framed windows on cold winter nights. Moisture condensing on windows is to be avoided because it can eventually lead to black mould spots forming on the glass.

To stick bubble wrap to windows, cut the bubble wrap to match the size of the glass. Lightly spray the window with water, then apply the bubble wrap. It should stick, simple as that.

A product that is slightly more sophisticated than bubble wrap and acts similarly to double-glazing are clingwrap-like films that are stuck to window frames using double-sided tape. The downside of these films is that they can't be reused seasonally. Therefore, they are likely to remain stuck to the window until they are damaged or degrade.

Other last-resort approaches for keeping a home warmer in winter may involve fixing a polyester cloth material (e.g. polar fleece) or cardboard directly to the glass. Of course, unlike bubble wrap, these methods won't let any light through.

External summer window shading

To be as comfortable as possible and to keep summer cooling costs manageable, it is critical to keep the intense Australian summer sun from directly hitting your windows. Likewise, exterior window coverings can also protect your windows from the sun's radiant heat coming from other directions, reflecting off wooden decks, brick walls or steel roofing. I've measured the temperature of decks in the midday sun as hot as 75 degrees Celsius, nearly hot enough to fry the proverbial egg.[2]

If they're all you have, interior window coverings are of some use in stopping heat from entering the home. However, it's something like ten times better to use exterior shading to stop the heat from reaching the home in the first place. Use both if you have them.

Properly sized eaves are a wonderful thing for protecting windows, not only from the sun but also from the damaging effects of rain on window frames. Sadly, I see more and more homes being built without properly sized eaves or any eaves at all.

There is an enormous range of exterior shading products on the market in Australia, ranging up to remarkable (but also

expensive) electrically driven self-erecting awnings that pull themselves back in under windy conditions. Cheaper options can include self-installed shade cloth.

To protect the windows at our home from the summer sun, we have some eaves, some shading plants, some pull-down canvas awnings and some cheap plastic exterior roller blinds. I'll also open up an umbrella to shade a deck area, although the umbrella and roller blinds can be unstable in high winds. If you're looking for exterior shading options, always consider how they'll perform in gusty conditions.

Aluminium Renshade can be a lifesaver

If you can't install exterior window protection for your home, Renshade – perforated aluminium foil on a paper backing – is a relatively cheap fix: renshade.com. To keep it out of the rain, Renshade is installed only on the inside of the window (like the bubble wrap described on page 156 for winter use) using Velcro stickers.

The perforations allow some light through, and you can actually see through them well enough to appreciate a distant horizon or to identify a person walking through your back gate.

At our house we apply Renshade seasonally on some upper floor windows that have no exterior protection other than eaves. We then save it and re-use it each year. Those windows are under attack from the direct sun and also from heat reflecting off steel roofing. Renshade certainly keeps our upstairs cooler, but I must admit from the outside it can make it look like I am running a meth lab.

Cardboard is another material that, when wedged against interior glass, will reduce heat coming in. But of course, cardboard will block all light.

Upgrading single-glazed windows

If you've done as much as you can with window coverings inside and out, your mind may turn to thoughts of upgrading your single-glazed windows, as well as any doors that have a lot of glass.[3]

There are number of ways to gain thermal improvement, such as:

* secondary glazing single-glazed windows by adding high- quality acrylic plastic
* retaining your existing window frames but replacing single-glazed glass with double-glazed glass
* refurbishing the existing window to repair any frame damage, improve opening and locking mechanisms, draught-seal and upgrade to double-glazed glass all at once

* adding a second window to the existing window, either within the interior window frame or attached to the outside of the home
* full replacement of frames and glass.

You can also use bubble wrap or Renshade, as I described before.

Unfortunately, improving windows can be expensive. Lower cost methods, such as attaching acrylic-plastic secondary glazing or knocking out single-pane glass and replacing it with double-glazed while retaining the existing frames, might cost $300 to $500 or more for every square metre. Fully replacing windows, frames and all, can cost from $1,000 to more than $2,500 per square metre.

Unlike some other home upgrades, such as heating with an air conditioner, improving roof space insulation or DIY draught-proofing, you probably won't make your money back on window upgrades via reduced energy bills for many, many years. So what are the reasons to upgrade windows? How can you prioritise which windows to address first?

Below is a list of reasons you might choose a particular window to upgrade:

* the frame is rotten or corroded
* moving parts and locks have failed
* an old window is leaky, perhaps never having had air seals in the first place, or the seals have deteriorated, or frames have warped, allowing more air leakage
* occupants would appreciate a significant reduction in street, weather or bird noise
* to make a window that you often sit next to on cold or hot days or evenings feel more comfortable
* to reduce moisture condensing on the glass on cold mornings

✱ a different opening method would be useful, for example, an easy to open casement window replacing an awning-type window that is hard to open; or sliding doors may be preferred over French-style doors

✱ aesthetic preferences.

Upgrading windows can certainly change a home. Though it won't result in a noticeable reduction in your energy bills, the improvements possible in thermal comfort, noise reduction, ease of opening, closing and locking, cleaning, maintenance and aesthetics can all be worthwhile.

At one stage, we had double-glazed windows in the front of the house but older, single-glazed windows out the back. Sitting in the front reading a book I'd find it nice and quiet. It was only on moving to the back of the house that I would notice if there was a windstorm raging. Such is the noise reduction possible with double-glazed (and well-sealed) modern windows.

How to rate and compare window upgrades

The various window upgrade methods described in this chapter will perform at different levels, providing different levels of improved thermal comfort, noise reduction and savings on energy bills. So how to rate and compare all the options?

Many readers will be familiar with the concept of the 'R' rating for insulation. When you go to the hardware store, you can see bundles of insulation with the R value written on the side: for example, R2 for some wall insulation or R4 to R6 for roof insulation. The letter R essentially stands for resistance to heat flow. With respect to reducing heat flow, the thicker the insulation the better, therefore the correspondingly higher R value, the better.

Windows are graded with the 'U' value, which is simply the mathematical reciprocal of R. This means for windows, the lower the U value, the better. Here is a range of (example only) U values for different types of glass in different types of frames.

Table 1

Glass type	U rating
single-glazed glass with ordinary aluminium frames	7
double-glazed glass with ordinary aluminium frames	4
double-glazed glass with uPVC (plastic) or timber frames	2
triple-glazed glass with uPVC frames	1

What this means is a window with a U value of 7 will lose 3.5 times more heat than a window with a U value of 2. If the window is large, as many of our windows are, the comfort improvement in both winter and summer will be noticeable. This will especially be the case if other aspects of the home's thermal envelope have already been improved, leaving the windows as the weakest link.

If you are looking at comparing window upgrade options, ask your supplier, 'What's the U value?' so you can compare the thermal performance of different options.

There are other technical parameters that can be used to score and compare windows, such as:

* the level of air infiltration
* solar heat gain coefficient
* visible light transmission.

 See Further reading and watching on page 200, and refer to the detailed information on the Windows Energy Rating Scheme website.

Secondary glazing with acrylic plastic

Secondary glazing with sturdy, high-quality acrylic plastic is a relatively non-invasive way to upgrade window performance. It is added to the existing window and can end up looking fine. I have seen this applied in quite 'high-end' homes. I am aware of two suppliers/installers in Australia: ecoGlaze and Magnetite. DIY techniques have also been described in past issues of *Renew Magazine*.

One application of this method is as follows: a spacer is applied at the edge of the existing window on the inside. High quality (stiff, 4.5 millimetre-thick) acrylic plastic is put in place to fit the window. A wooden or plastic frame then holds it all together with adhesive/magnetic strips. The air gap that is formed is similar to the gap that exists with a double-glazed window, with the trapped air between the glass and acrylic plastic acting as insulation.

Retaining your window frames while upgrading to double-glazing

Another window upgrade method involves retaining the existing window frames (or potentially restoring them somewhat if there is damage) and replacing the existing single-glazed glass with a sealed double-glazed glass.

Many glaziers can offer this service, especially for timber-framed windows. I know of one specialty company doing this: Retrofit Double Glaze (formerly known as DIY Double Glaze). Also, the vendor TwinGlaze can help with aluminium-framed windows.

Note that not all double-glazed windows are created equally. Ask about the U value for the glass being supplied. Double-glazed glass can be twice as heavy and obviously thicker than the single pane of glass being replaced. This means there may be limits as to what can actually be done with existing windows from the point of view of allowable weight and dimensions.

With a view to minimising physical dimensions and weight, at least one Australian vendor supplies a double-glazed window with a vacuum gap between the two panes, instead of filling the gap with argon gas, as is done with most double-glazed windows. This can provide good thermal performance (low U value) because this glazing is constructed to act like a vacuum thermos bottle.

Refurbishing existing windows

Vendors such as Thermawood, SealaSash and others will fully restore a window, upgrading to double-glazed and including updated modern draught seals. This can be a handy service if the household wishes to retain the heritage characteristics of, for example, poorly operating sash/double-hung or other styles of windows and doors.

Adding a second window, inside or outside

Vendors such as Stop Noise can add a second window inside your existing window if the existing window has suitable frame depth. This is most often done, as the company name suggests, to reduce noise coming through the window. Improved thermal performance will be a side benefit. A drawback is that in order to open the window to let in air, you must open two windows.

Where I grew up in the US and before double-glazed windows were available, a second window was applied to the outside of a home. We called these storm windows. The glass in the window would be seasonally replaced with flyscreens for summer and the glass panes stored away until the next winter.

Replacing your windows with double-glazing

I recommend that any new window you buy in Australia should be double-glazed. I mean, if you go to the shops to buy a new phone, you're probably going to end up buying a mobile and not a landline phone. The technology has moved on with phones, as it has with windows.

If you are in the market for new windows, get down to the double-glazed window showroom and fall in love with windows. Think of it like buying a new car. Indeed, what you're going to spend might be similar.

Discuss with the salesperson what sort of opening configuration you want for your doors and windows. Sliders or casements? Stacking or French? In the showroom, feel the operation of the doors, the tightness of the seals and latching mechanisms.

Remember, not all double-glazed windows are created equally

As I've mentioned, not all double-glazed windows (including the frames) are created equally. To continue the analogy of buying a car: you might be thinking of buying a car, but what sort? Not all cars are created equally.

Most often, I recommend that people strongly consider uPVC plastic frames. Because plastic is a reasonable insulator, these frames can provide excellent thermal performance similar to timber but require far less maintenance. You'll never need to paint uPVC frames. They are also generally more affordable.

If you are considering aluminium frames, ensure that they are 'thermally broken'. This means there will be a piece of insulating plastic within the aluminium frames that acts as insulation and 'breaks' the transfer of heat through the window frame.

Good double-glazed glass will reduce the chance of water condensing on your windows in winter. However, if your windows have cheap aluminium frames that are not thermally broken, you may see water condensing on the cold unprotected frames themselves. Well-performing thermally broken aluminium frames are likely to be more expensive than uPVC frames, so this takes us back to my recommendation that uPVC frames could be your best option (see page 162).

And then there is the glass itself. There are a lot of variables when it comes to the construction of double-glazed windows, so there are a lot of questions to ask, such as:

* How thick is the glass?
* How wide is the spacing between the two panes of glass?
* What gas fills the gap? Is it argon?
* Has a 'low-e' coating been applied (see below) or have any other tints?

The best way to evaluate the many window products is to compare the U values, as well as the solar heat gain coefficient, the visible light transmission and other parameters.

Should you consider going beyond double-glazed and buying triple-glazed windows? My general view is that you can be comfortable and have low energy bills and a low environmental footprint with just double-glazed windows in Australia but, that said, if a supplier is offering triple-glazed windows for not more than the price of double-glazed, why not? Note that triple-glazed windows will be heavier and the frame thicker than with double-glazed.

Add low-emissivity (low-e) coating

I generally recommend that a low-emissivity coating (aka 'low-e' coating) be applied to new double-glazed windows. This is a metallised coating that you can't really see, which goes on the inside of one of the panes of glass. The coating helps to reduce heat transfer in winter and summer. It has a positive effect on the U value of the window. It isn't something that you can add later so you might as well have it from the beginning.

It's not technically correct for me to refer to it as sunscreen for the windows, but that analogy might help you to remember to ask for low-e coating when you are buying new windows.

Our homes have too much glass

Try not to overdo it with windows. Even if you were to install a triple-glazed window with low emissivity coating and excellent coverings inside and out, this window will still be your home's thermal weak link. Windows are costly, both in terms of money and potential for energy loss, so please have a serious think about every square centimetre of glass that you are purchasing.[4] Could you get by with less?

For example, in our television room, formerly we had 10 square metres of single-glazed glass. During a renovation, we replaced that with 5 square metres of double-glazed glass. Formerly the windows went right down to the ground, even though there was no need for this feature. With this renovation, we now lose only one-sixth of the amount of energy out through the windows. It's been a game changer for the comfort of that room. We don't feel that we're missing out by having less glass, and we feel cosier and more comfortable. We even saved some money on our new blinds, given that our windows were smaller than they might have been.

Be wary of skylights

Many Australian homes have skylights. Architects and designers and the folks on *The Block* love to add more and more. But I am wary of skylights. Why? Here are a few reasons:

✶ A skylight is a hole in your roof and could leak rainwater.
✶ A skylight can be damaged by hail.

* A skylight can develop lichen and moss on the outside and spider webs on the inside and might not look that great a few years down the track.
* The skylight, along with its accompanying uninsulated tunnel/shaft, can let in uncontrolled heat in summer and let heat out in winter. In this way, a skylight performs badly in comparison to an insulated section of ceiling/roof.
* Skylights often get in the way of a home's solar PV panel layout.
* Skylights are expensive, whereas these days a wide range of simple and far lower-cost LED light panels are available that can provide better service. LED light panels are simply installed on your ceiling like other common light fixtures. These can even fool your friends into thinking you've just installed a skylight, except that it will also work on dim days and at night! LED light panels can be fitted with light detectors and motion sensors, providing the light you need when you want it, while not allowing in any extra summer heat.

12.
Minimising electricity use and costs

As you start electrifying your home, it's good to monitor how much electricity your home and/or car charger is actually using, at what times of day and how much you are paying for it. You might ask yourself:

* How much electricity does any given appliance use?
* How can you minimise electricity use?
* In homes equipped with solar PV panels, how can your electricity use be shifted to sunlight hours, when the cost is cheapest?
* Which electricity retailer offers the best electricity supply deal? Will it be a time-of-use tariff or a fixed rate?
* How does one read an electricity bill anyway?

In this chapter, I discuss some of those questions and point you to further resources.

What is a smart electricity meter and why would you want one?

Let's start with your electricity meter, which is on your property but is owned by your electricity distributor. A smart meter is generally a meter that wirelessly beams information back to electricity company headquarters all through the day. Therefore,

smart electricity meters don't have to be read by a meter reader. Victoria has had these since the roll-out began back in 2006.[1] Since then, some other states and territories have followed suit.

One benefit to you of a smart meter is that you can access data describing your electricity purchases and any solar PV–generated sales too, every few seconds in fact. Why not, it's your data! You may be surprised to discover what this data says about your home. With a smart meter, gone are the days when, every three months, you'd be shocked to learn how much electricity you had used.

Accessing your smart meter data

Once your home has a smart meter, you should be able to access your data by logging onto the online web portal managed by your electricity retailer or by your electricity distributor, or both. There you will find access to a file of half-hourly data (purchases and solar PV exports) that you can download, as well as charts and graphs already prepared by your retailer or distributor.[2]

This data gets uploaded to these online portals only once a day, so to get the best understanding of what's going on in your home, occasionally keep a log throughout a day as to what appliances you are using. Then the next day, you can compare your log to the data that appears on the online portal. Can you explain any big spikes in electricity use? How low can you go with your electricity use in the middle of the night? What difference does it make to turn off the NBN modem or that old refrigerator in the garage? How does your home's consumption change as the seasons change?

Note that in these portals, the most granular data you will see is half hourly and is uploaded a day later. You won't see what your electricity use is right now. To access a nearly instantaneous reading of what electricity you are purchasing/using right now, you can buy a home electricity monitor. This allows you to access your purchase data via your phone on a nearly instantaneous basis. A few government programs and some electricity suppliers have offered these for free or at low cost.[3]

A home electricity monitor should provide charts of half hourly, daily or monthly electricity purchases. Note that the simplest of these devices only report your electricity purchases. If you have solar PV panels on the roof and if at any time you aren't buying electricity but selling it back to the grid, it will just read zero at those times. Nevertheless, even this information can be useful because it can be comforting to know that your own electricity generation is covering your home's needs and maybe even more.

In the homes of my clients who happen to already have an electricity monitor, we go on a little hunt. We shut off electrical appliances and devices one by one, noting how much electricity each one was using, until we get down to a very low level, say only 25 watts of electricity being used at that moment. Householders are often surprised to find out how much electricity an electric-resistive heater uses compared with, for example, a reverse-cycle air conditioner. They may also be surprised to see how much electricity some appliances use, such as gaming computers or old 'vented' clothes dryers.

Monitor consumption with solar PV

If you're thinking of installing solar PV panels on your roof, enquire also about getting home electricity 'consumption monitoring' installed. This will cost extra for the equipment involved, but it will give you access to information about the following things and how this all works together:

✴ how much electricity your home's solar system is producing
✴ how much electricity your home is consuming
✴ the state of a home battery or electric vehicle.[4]

If you already have solar PV panels on your roof but no consumption monitoring equipment, you can buy retrofit equipment.

 For information on solar PV panels, you could try Solar Quotes (solarquotes.com.au/good-solar-guide/monitoring-systems). For retrofitting consumption monitoring equipment, Solar Analytics (solaranalytics.com.au) should be able to help.

How much electricity do individual appliances use?

These days many councils or libraries are loaning out home energy kits. Often the kit comes with a power meter, a device that can be used to assess the power and energy demands of individual appliances. Plug an appliance into the power meter and then plug the power meter into the wall and away you go.

It can be especially enlightening to plug a refrigerator into the power meter and let it run for a week. After the week has elapsed, you'll be able to see how much money you spent running your refrigerator.

Here's a sad story about refrigerators. When you buy a new fridge, you might think it's wasteful to throw the old one away or even to see it go off to the recyclers. Perhaps you're thinking you'll throw a party someday and might need to keep some extra drinks cold. So this old refrigerator, which is nowhere near as well insulated and efficient as the new one, ends up in a hot garage, next to a hot car. And then you expect it to keep a few drinks cold without costing a lot of money?

Be warned, it will cost hundreds of dollars each year to run that fridge. I joke that in this situation you should take the new fridge, which is more efficient and capable than the old one, and put it out in the hot garage, while keeping the old fridge right there in the kitchen where it always was! But here's a better idea. Don't have any more refrigerators in your home than you really need, and don't send an old refrigerator out into the garage to struggle.

Don't have any more refrigerators in your home than you really need.

Where can you look for other electricity waste around your home? Anything that feels warm to the touch is probably wasting electricity. Old plasma televisions were huge energy wasters and were hot to the touch. Today's big screen TVs can use a lot less electricity and, because of that, they don't even feel that warm. We used to have a small clock radio in the bedroom. It was always warm to the touch. Heat from things that aren't meant to be heaters is an indication of wasted electricity. The clock radio was wasting about 40 watts of electricity, so we stopped using it. Of course, these days, you can use your phone as your alarm clock if you want to.

You can find the government-regulated energy star ratings for many electrical appliances available in Australia at energyrating. gov.au. It's best to have reviewed the information at that website before heading out to the appliance shops. At the shop, you might see an appliance that's rated 4 stars. While that might look pretty good, you might find out later that a more efficient 7-star appliance was available somewhere else.

Getting the best electricity supply deal

You've heard this advice already: shop around to get the best electricity deal. There are comparison websites set up by state and federal governments,[5] and also by businesses such as WATTever.

When you buy and sell electricity, there are a few different parts to what you pay or get paid for. Households with large solar PV systems on their roof may be attracted to a supply deal that advertises the highest feed-in tariff. My caution is don't ignore the rest of the deal, which includes what you pay for the electricity you import, and also the daily supply charge.

Furthermore, there is no set and forget with gas and electricity bills these days. You have to keep an eye on them. I suppose it's a bit like booking a plane flight – the prices can change every time you look at them. Also, consider the possibility that your electricity or gas supplier is overcharging you:

* because they can, and
* because they wouldn't mind too much if they lost you as a customer.

There may be something their algorithm doesn't like about your supply and demand profile. Perhaps the only way they want you to remain with them is if you pay exorbitantly high fees.

Read the messages the energy suppliers send you. Have a look at the fine print spread across the bill. It might even say right on the bill: 'This isn't the best deal you could have.' I guess they're required by law to have that printed there, but if you're not watching, you'll end up paying too much.

Your household may be on a flat electricity rate (or tariff), paying just one set price for each kilowatt-hour of electricity you buy. Or you may be on a time-of-use tariff where you pay one price at certain times of the day or week and a different price at other times. But which is best for you? If you've been in your home for at least a year, and if nothing significant is changing about your home over the next year, and you have access to the data from your smart electricity meter, you or your electricity supplier should be able to work out which is the best tariff for your home.

I've seen homes on time-of-use tariffs for no good reason that I could see. In fact, some householders may not even know what tariff they are on or that there are indeed different tariffs available.

Time-of-use tariffs can be useful if you have, for example, a pool pump that you can run during certain low-price periods, or perhaps you can charge your car or a home battery or heat your water at such times. But some home occupiers don't have such things and aren't able to shift their electricity use around throughout the day. They may be better off on a single price, flat tariff.

Access to wholesale prices with electricity demand-response deals

There are even more sophisticated deals out there for electricity users who are far savvier than I am.

Amber is one electricity supplier that seeks to give their customers access to wholesale prices, prices which are often low but are sometimes scarily high. Wholesale electricity prices in eastern Australia can range from minus $400 per megawatt-hour to an extremely high $20,000 per megawatt-hour.[6] (To compare with retail prices, these convert to a low of minus $0.40 per kilowatt-hour and a high of $20 per kilowatt-hour.)

Ideally, households accessing low wholesale prices can use them to, for example, charge an electric car. The flipside is that you'll want to stop using electricity during the high price periods. I've heard of Amber customers pulling cakes out of the oven half-baked. This sounds inconvenient, but they didn't mind because over the long run they thought they were onto a good thing.

We'll see more and more of this sort of demand–response activity going on.[7] Some homes may be set up so pool pumps, electric-resistive water heaters, air conditioners or car charging automatically pauses during times of high wholesale electricity prices.

Other resources to help lower your electricity bills

Many government bodies and sustainability oriented not-for-profit organisations offer tips on ways to reduce household energy bills.[8]

SEE ALSO *Further reading and watching (page 200) for specific resources*

13.
Solar PV panels and batteries: are they worth it?

With around one-third of Australian homes now having solar PV panels on their rooftops, just about everyone already has them or knows someone who does.[1] The average size being installed is approaching 10 kilowatts, which is more than twenty panels. What a difference compared to when we installed six small panels on our home in 2008, or even some years before that when the true early adopters invested in solar PV at a cost ten times greater than today.

We're also hearing more and more about home batteries, batteries in cars and even community batteries. But what can you do with solar PV and/or batteries? Are they worth getting for your home? And what if you don't live in a free-standing home? How can you benefit from these technologies?

These are some of the questions I cover in this chapter.

Should you install solar PV panels on your roof?

If you don't already have solar PV on your roof, should you get it?

In 2016, I wrote an article with the cheeky title: '22 ways to cut your energy bills (before spending on solar PV panels)'.[2] My point was that although solar PV was popular in 2016 and still is, there may be better home investments. This is the reason I've placed the subject of solar PV toward the end of this book. You might achieve greater value by heating with an air conditioner, by draught-proofing or by any of the other suggestions I've described up until now.

That said, if you don't already have solar PV and unless your roof is heavily shaded, I suggest you get some designs and quotes as soon as you can. Assess how much electricity you can generate and how much you may be able to save on your electricity bill. At no cost to you and with no visit to your home required, solar PV suppliers use aerial mapping tools to quickly sketch up a design for panels on your roof. With this comes an initial estimate for what the system will cost, with and without rebates. Critically, they'll also provide an estimate of how much electricity your system will produce over the course of the year and month by month, including the critical winter months.

Reduce your electricity purchases with solar PV

You can work out some rough economics by firstly assuming your home will use (only) about 30 per cent of the electricity you generate. This means you'll be buying less electricity from the grid. It's known as 30 per cent 'self-consumption' or 'own consumption'.

Every home, along with its occupants, is different, however 30 per cent is a rule of thumb based on what I have seen happening for a range of homes in various locations, including my own. However, these homes don't have swimming pools, electric vehicles or home batteries. Those items can boost self-consumption to 40 or even 50 per cent but probably not much higher than that.

Feeding into the electricity grid

The other 70 per cent of what your solar PV system produces you will sell to the grid. Electricity feed-in tariffs vary across the different states and territories, but these days they are quite low and falling. Why? Often on sunny days, there is so much rooftop solar PV electricity available that wholesale electricity prices plunge into negative territory.

You will need to check the tariffs for your particular location. Using my home state of Victoria, for example, you might assume that you can sell your electricity only at a price as low as $0.03 per kilowatt-hour.[3] Long gone are the days when a premium feed-in tariff of $0.60 per kilowatt-hour prevailed. You may wish to check the economic payback of a new solar PV system and, assuming a worst-case scenario of $0.00 per kilowatt-hour feed-in tariff, that means you'll be giving any exported electricity away for free.

Furthermore, you might also have to consider export constraints. Will your electricity distributor limit what your system can export out to the grid at any time to, say, only 5 kilowatts or perhaps even as little as zero kilowatts? Why do electricity distributors set these limits? There may be a few hours each year, such as at midday two days after Christmas, when so much solar electricity is being produced in your neighbourhood that the electricity distribution grid has difficulty handling all of it. The risk is that too much solar generation can lead to voltages being too high on the system.

Unfortunately, electricity distributors setting a constant 'dumb' limit of, say, 5 kilowatts, is taking, as they say, a sledgehammer to the problem of cracking a nut. Hopefully we'll see more sophisticated approaches to managing solar-generated electricity that constrain production for only those few hours a year when it's necessary to do so, and that don't penalise households all year long.[4]

Solar PV financial payback

Continuing with your investigation into whether solar PV is a good investment for you, you can now add the value of the electricity you didn't have to buy to the value of the electricity you sold out to the grid, and divide that sum into what you'll have to pay for equipment and installation. This gives you the payback period, or the length of time it will take for you to recoup the cost of buying and installing the PV system.

Here's an example calculation. Let's assume you have a quote for a solar PV system that has a capacity of 13 kilowatts, that is going to cost you $15,000 and is estimated to produce 14,000 kilowatt-hours in the first year. And let's assume it costs you $0.30 per kilowatt-hour for electricity you buy from the grid, and that the feed-in tariff you'll receive for electricity you sell is only going to be $0.03 per kilowatt-hour.

Multiply the annual production by the 30 per cent self-consumption and your electricity price means you might reduce your electricity purchases by $1260 in the first year.

See below:

14,000 kilowatt-hours × 30% self-consumption × $0.30 per kilowatt-hour of electricity you didn't have to buy = $1,260 saved a year.

The value of the electricity you sold will be $294 in the first year:

14,000 kilowatt-hours × 70% exported × $0.03 per kilowatt-hour = $294.

When we add the value of the electricity you didn't have to buy with what you were paid for selling electricity, we see you gain:

$1,260 + $294 = $1,554 a year.

Dividing this into the purchase price gives us a payback period of 9.7 years:

$15,000 $1,554 = 9.7 years.

These days, I'm often seeing payback periods of eight or more years, but of course this will vary home by home. Payback periods are lengthening as feed-in tariffs approach zero. Electricity prices vary greatly across Australia. You may be able to reproduce the above calculation using your own data and assumptions.

Is an eight-year (or longer) payback a good enough return to justify the investment? It means up until year eight the reduction in your electricity bill and any feed-in benefit you receive will have gone to covering what you paid for the system. At that point you won't have made a positive financial return. But then in year nine you may start to profit by one or two thousand dollars every year after that, for as long as your solar PV system lasts.

How long do they last? Our sixteen-year-old system is still chugging along, as are the sixteen-year-old systems owned by my neighbours who got involved in the same bulk-buy. At least one set of early solar PV panels on a home in Vermont, US, is still working after 40 years. Warranties can range up to five or more years for the inverter equipment and up to 25 years for the solar panels.

If the payback period you calculate is around eight years, it's not a no-brainer to get solar, but it can still be a more attractive investment than having a term deposit in the bank. And even if investing in solar means adding more onto your mortgage, it may not be the silliest thing you could do with your money.

Government-supported rebates and credits tend to get smaller every year. I encourage people to look into getting solar PV sooner rather than later, and then either do it or forget about it.

Accelerate your solar PV payback by electrifying everything

Please note that some of the assumptions I made in the previous calculation could be wrong. Maybe your home and vehicles won't consume only 30 per cent of the electricity generated. The level of self-consumption will be higher for smaller solar PV systems versus very large ones.

Or perhaps you have a swimming pool, a very large household or you will be getting a home battery or an electric car or two. All those things could mean your household has or will have high electricity usage. This could push your self-consumption level up to 40 or 50 per cent, improve your financial return and reduce the payback period.

In the calculation I did, assuming a self-consumption level as high as 50 per cent instead of only 30 per cent drops the payback period from 9.7 years to only 6.5 years. That's a big difference.

Fully electrifying your home and vehicles is a winning strategy to get the most out of your solar PV system. Do all of the things I described earlier in this book about getting off gas and electrifying your living-space heating, hot water and cooking.

Electrify your lawn mower (and other home tools)! While I wait to someday buy an electric car, for now I'll make do with our little electric Ryobi push mower. Actually, the best thing about it is that after a break of about 30 years, once again my wife is happy to mow the grass. Thanks Ryobi mower! Thanks electricity!

Of course, to benefit from your solar-generated electricity, those electrical things will need to be powered during sunlight hours, unless you have a home battery or can power your home off the battery in your car. (See more on that on page 188.)

Importantly, in terms of investing in our environment and helping to stabilise our climate, in most places in Australia installing more and more solar PV panels on rooftops continues to help reduce the burning of fossil fuels to make electricity, especially in winter.

Go once, go big?

An obvious question is, if you are going to install a solar PV system on your roof, how big should it be?

When you're at the stage of getting various designs and quotes, some of those designs should include as many panels as you can fit on the logical places of your roof, by which I mean, for the Southern Hemisphere, the unshaded east-, north- and west-facing parts of the roof. North is the most productive direction, but east- and west-facing will also be useful. Flat roofs will require racks to tilt the panels up at an angle in a preferably north-facing direction. Some Australian homes have even installed south-facing panels. These will generate useful amounts of electricity in summer, but not so much in winter.

After getting your quotes for as much as will fit on the logical parts of your roof, consider what such a system will cost versus the value of the electricity produced, and of course what you have to spend or can get financing for.

At the Facebook group MEEH, with more than 116,000 members, I'm yet to hear a member express regret that their solar PV system is too large. Whereas, particularly in winter, we often see many members thinking about installing a second, smaller solar system on areas of the roof that they still have free. Unfortunately, this 'top-up-later' option usually has a poor financial return. It would have been better to have installed a larger system in the first place.

In recent times, it has become customary for solar vendors to offer quotes for a 6.6-kilowatt system. I call this a cookie-cutter size that the vendor has come up with, giving no great thought to your needs or what your roof can handle. Each home is special and unique. Avoid cookie-cutter quotes.

When you are considering investing in solar PV, check what the limits may be on what you can install or how you can use it, as set by regulators or distribution grid owners.[5]

Think winter!

When finalising your sizing decision, focus on what your solar PV system will produce in mid-winter. It's common to hear people brag about what their system produces in the summer months, but mid-winter is when we really need electricity. On cloudy winter days, nearly everyone's solar PV system is too small. So when deciding how large your solar system will be, think mid-winter!

Indeed, when getting quotes, compare what the different design options will produce in the month of June. This is when we need as much electricity as we can get to heat our homes, heat our water and charge our cars.

Though at our home, our annual average self-consumption is only 30 per cent, it ranges from a high of 54 per cent in June to a low of 22 per cent in December. This is one way to show how valuable solar PV electricity can be in the winter months.

Should you get a home battery?

When I'm speaking with people in their homes or at community meetings, anyone who doesn't already have a home battery will ask me: 'Should I get one?'

Does your home suffer from recurring power outages due to floods or windstorms knocking down branches and trees onto power lines? If so, then you may wish to wire up a solar PV system along with a battery to act as a back-up power supply for your refrigerator, lights and computers – appliances that consume only modest amounts of electricity. In this situation, you'll want to keep your battery charged up in advance of likely storms or other disruptions. (Below I describe how an electric vehicle may someday provide the same service, provided it's parked at your home.) In the case of very long outages, an old-fashioned fossil fuel-burning generator might end up being a necessary option.

Or are you an early adopter who wants to learn what a home battery is all about? In that case, get one! To gain the most value out of your battery, you might wish to sign up with an electricity supplier such as Amber that provides access to wholesale prices. As I described earlier, these prices can sometimes be negative, meaning that you can be paid to use electricity, but at other times can be excruciatingly high. The aim would be to charge up your battery when the electricity price available to you is cheap, free or even negative, and then have the battery supply your home's electricity requirements in the evening hours to avoid importing electricity when prices are high. If you are an enthusiastic battery adopter and willing to participate in this, a battery might actually be a better investment than a solar PV system.

But for most households, generally my response to the question of whether they get a battery is, 'No, not yet.' Why not?

How a battery is like a rainwater tank

Home batteries generally don't provide a positive financial return, given what you have to pay for one (ranging from $8,000 to $15,000 depending on battery size). At current economics, the payback period can be twenty or even 30 years, far beyond a battery's standard ten-year warranty period. I can usually identify a long list of better ways for people to spend money improving their homes.

In a way, electricity-storing batteries are like the rainwater collection tanks that were installed at many of our homes during periods of drought. It may be that the tanks indeed reduced water consumption, and so people who invested in tanks thought this was a good outcome. However, what's not clear is whether households that installed a rainwater tank, or the governments that subsidised them, saw a good financial return on their investment.

Please allow me to keep going with the battery/rainwater tank analogy for a moment. To get the most out of a rainwater tank, you need three things: a sizable roof to collect the water, rain to come down and fill up the tank and a reason to promptly empty the tank so that the next time it rains you can collect more rainwater and go through the cycle again. If your home has a small roof, if it doesn't rain or if you don't have a recurring need to use the water, a rainwater tank might be a poor investment, especially when compared with some other home upgrade.

I remember when a friend was very excited about the rainwater tanks he was installing at a cost of thousands of dollars. It eventually rained and I imagined he would be happy that his tanks were now full. But no, the next time I saw him he was looking sad. He had worked out that after spending a lot of money on the tanks, the water contained in them was worth only two dollars.

It can be a similar thing with home batteries.

With current electricity tariffs and battery technology, there may be only two dollars' worth of value that can be held in a home battery. That's the value to be gained by charging your battery with cheap or even free electricity, and then using it to power your home at night when electricity prices are high. Over a given day, if you manage to fill the battery and empty it again, you might save two dollars on your electricity bill.

To cycle the battery and make that two dollars, you'll need to have enough sunshine and solar PV panels on your roof to generate the electricity needed to run your home and charge your electric vehicle during the day before you can start to charge your home battery with what's left over. Then you'll need a reason to empty the battery that night so that the whole cycle can start again tomorrow.

For many homes, there won't be many days a year when it works out like this.

> There may be only two dollars' worth of
> value that can be held in a home battery.

Let's be optimistic and assume you did manage to cycle your battery 300 days a year and gain two dollars of value each time. That's $600. If you did this consistently for ten years, that's $6,000. The problem is, by then the warranty will have expired, the battery performance will have degraded somewhat (not unlike a phone or laptop battery) and you are still trying to gain a value equal to what you paid for it.

However, as I wrote previously, for those enthusiastic about participating in the wholesale market, some Amber customers have reported earning not two dollars a day but rather as much as $32 a day. Such days might not come along that often, but participating in wholesale pricing is one way to improve the financial return of a home battery.

Keep in mind that when you put electricity into a battery, you may only get about 90 per cent of that back. Batteries have a 'round trip efficiency' of less than 100 per cent. Ten per cent might be degraded to waste heat. In this way, batteries are like a leaky rainwater tank.

Home batteries need to get cheaper

A few years ago, it was projected that home batteries would become cheap enough that they would be a good financial proposition for home owners. This hasn't happened yet. Why not?

One reason that home batteries have not come down in price is that everything around the battery, referred to as the 'balance of system', tends to get more expensive over time. For example, increasingly stringent safety standards have increased the cost of installing a home battery.

Another reason is that much of the lithium produced in the world is going into batteries for electric vehicles. Lithium is a light material that makes sense to use in vehicles where weight is critical. But it doesn't make sense that home batteries are trying to compete with vehicles for light lithium. We need a cheap alternative chemistry to emerge that doesn't involve lithium.

But don't despair. I foresee that many households will be getting a battery soon, and it's going to be a big one that comes with four wheels!

Your electric car may power your home

What I'm seeing in clients' homes is that the first big battery they are investing in is in their electric vehicle. An electric car's battery will be six or seven times larger than any home battery would be. And someday, provided you install the necessary equipment, the battery in your car will be able to power your home.

Trials of this, known as Vehicle-to-Home (V2H) or Vehicle-to-Grid (V2G) operation, are being conducted in Australia and around the world. In future, possibly as early as 2025 in Australia, the electrical energy stored in the millions of our vehicle batteries could become an important component of Australia's electricity supply system.[6]

Even now during an emergency, you can use the battery in an electric vehicle to power 12-volt appliances, such as a camping refrigerator or lights. Going further, some electric vehicles, such as the BYD Atto 3, have what is known as Vehicle-to-Load (V2L) capability. In an emergency, some normal 220–240-volt household appliances could be run off this car.

Confused by all the acronyms? Here's one more: 'V2X' means when we can plug our electric cars into whatever!

Not in a free-standing house?

Solar PV systems continue to be extensively and fairly easily deployed on free-standing Australian houses. We may well lead the world in this. However, millions of Australians don't live in free-standing homes. How can people living in units and apartments benefit from solar PV?

One way that all Australian households connected to the electricity grid benefit is through the lower electricity prices that result from deploying renewable energy. A very large part of this comes from the millions of solar PV panels on individual rooftops, as well as utility scale wind and solar farms.

A good example of how renewable energy leads to lower electricity prices is the way that the residents of the Australian Capital Territory, where the entire electricity supply is sourced from 100 per cent renewable energy, did not see electricity prices go up as a result of the conflict in Ukraine that raised fossil fuel and electricity prices globally. As a result of this cost-saving policy, all ACT residents benefited, whether they owned their homes, were renters, lived in free-standing homes or apartments, and whether their homes had solar PV on their roofs or not.

For people who live in apartment buildings with a roof suitable for solar PV panels, the Australian company Allume Energy has developed and deployed innovative hardware and software, named SolShare, that enables solar-generated electricity to be shared from a single rooftop solar PV system to multiple dwellings in the same building. Here is one example: a SolShare project in Highett, Victoria, distributes electricity from a single solar-PV system on the roof of the building to five apartments, a baker, a hair salon and an occupational therapist housed below.[7]

 For more information on the world of solar PV and batteries, see Further reading and watching on page 200. You'll also find useful information at solarquotes.com.au.

14.
I'm a renter, what can I do?

Many of the initiatives I describe in this book are actions renters can take. In this chapter, I bring many of those ideas to one place.

 A great resource is the 'Coping Cookbook: Renters' Recipes for Resilience' which can be found on betterrenting.org.au.

Heating and cooling

Some rentals will have multiple ways to heat the home. The cheapest way can be with a reverse-cycle air conditioner, as described in Chapter 5 (page 61). This can operate at one-third the cost of gas-fired heating or one-quarter to one-fifth the cost of using a resistive-electric heater (e.g. a panel, fan or oil column heater).

Air conditioners, ducted gas heaters and many devices that blow air around will have filters. These can often be cleaned DIY. It is essential that they are kept clean both for the health of the equipment and for the health of the home occupants.

If no reverse-cycle air conditioner exists at a property, renters may be able to lobby their landlord to install one. Health studies have shown there are good reasons for people to have access to active cooling (air conditioning) during summer heat waves in order to improve health and avoid premature death. Governments are also encouraging landlords to make efficient and lower cost heating available to their tenants.

By rental property regulation, gas equipment needs to be professionally safety checked every two years to ensure tenants won't fall victim to carbon monoxide poisoning. Air conditioners do not require such a check. By switching away from gas, it may be attractive to a landlord to never again have to pay for a gas safety check.

It may come to pass that government policies promoting electrification, such as attractive tax depreciation, will be applied where landlords replace gas-fired heating with reverse-cycle air conditioning.

To minimise heating and cooling costs, a renter may be able to zone off different living areas by hanging a light sheet of cloth (known in Japan as a 'noren'), as described in Chapter 5 (see page 89). Renters can also view an article called 'Bubble wrap insulation and DIY air-con: how renters can keep their home cool during summer' in *The Guardian Australia*.[1]

SEE ALSO *Heating and cooling (page 61)*

Showers and water heating

Check your shower head. Renters may be interested in grabbing a bucket and finding out how much water their shower head passes. Ideally, a shower head doesn't pass more than 9 litres of water a minute. In some Australian states and territories, low-flow shower heads are being distributed and installed at no cost, although this may require landlord approval. Alternatively, a renter may be able to fit a flow restriction orifice to reduce shower head flow, as described in Chapter 6 (see page 92).

When moving from one rental property to another, a family member used to take their own low-flow shower head with them.

It was sometimes possible, without damaging any hardware, to switch on the low-flow shower head when they were living there, and then to switch the landlord's shower head back on as they were moving out.

Use a timer to access lower electricity prices. With an electric-resistive water heater or a hot-water heat pump, it may be possible to set a timer so that water is heated when electricity prices are cheapest. A timer may already be in place at a property, in which case the renter should investigate how to set it to heat water at the cheapest times.

Upgrade the hot water service. Dates of manufacture can sometimes be found on a hot water system. If a hot water system is more than ten years old, or even if it's not that old, a renter may be able to lobby a landlord to install an efficient hot-water heat pump. To the landlord, the cost to replace may not be high given the rebates and incentives available, including tax depreciation.

SEE ALSO *The cheapest way to heat water (page 93)*

Cooking

A renter may opt to immediately buy, plug into the wall and use a portable induction cooktop – or even two or three – if that fits their cooking and kitchen needs. If it can be accessed, shut off the gas valve to the gas cooktop to make it safe. The gas valve can be turned back on again when you leave.

For best health and home cleanliness outcomes, keep the grease filters in rangehoods clean. Use the rangehood fan at all times when cooking. Crack open a nearby window when cooking to allow the fan to work better.

SEE ALSO *Bye-bye, gas grid: cooking without gas (page 103) for the health benefits of using an electric-induction cooktop instead of gas*

No more gas – no more gas bill!

A renter may be receiving a gas bill. However, by using an induction cooktop and/or a reverse-cycle air conditioner (if available), a renter might suddenly find they no longer use any gas. In this situation, ring up the gas supplier and tell them to stop sending you a gas bill.

Draught-proofing and keeping your air clean

A renter may be able to do some minor draught-proofing that a landlord is unlikely to ever notice or object to, such as caulking up a gap behind a window architrave.

Some draught-proofing actions can be non-permanent and reversible, such as installing a chimney draught-stopper.

Some states, such as Victoria, are offering free draught-proofing items, such as bathroom fan dampers and door seals, although installation of these may require landlord approval.

What if it is difficult for a renter to improve a draughty room such as a toilet, bathroom or laundry? If the room has a door, cheap (less than $10) twin draught-stoppers – or what I refer to as 'double-door snakes' – can be bought and simply slid beneath the door. Then close the door to isolate these draughty rooms from the rest.

However, as I wrote in Chapter 9 (page 127), as we reduce the draughtiness of our home, we need to focus on keeping the air in our home clean. Renters may wish to procure air quality monitors to measure things such as humidity and small particulates.

SEE ALSO *Draught-proofing and clearing the air (page 127) for details on air quality management techniques*

Insulation

Improving roof space insulation can be one of the most cost-effective ways to thermally improve any home. If safe to do so, a renter may be able to open the roof space access manhole (if one exists) to take some photos of the insulation up there. Or perhaps find a friend or family member who could do this. If the insulation is old and thin, patchy and disturbed, or totally absent, you have a good reason to tell the landlord that it should be improved.

Some councils and libraries loan out infrared thermal-imaging cameras. Under the right weather conditions, it's possible to use thermal imaging to show that the roof space insulation is not consistent. This can be handy if a visual inspection of the roof space is difficult or can't be done.

SEE ALSO *Insulation: a priority (page 144)*

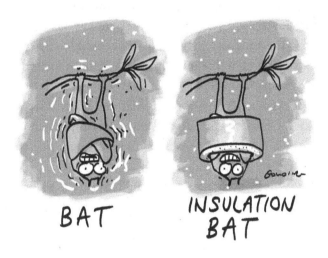

BAT

INSULATION
BAT

Windows and window coverings

In Chapter 11, I wrote about how bubble wrap can be used to achieve a double-glazing effect for little cost and in a reversible way (see page 156). By reversible, I mean the bubble wrap can be installed without damage to the property and can be removed without a trace seasonally or when the tenant vacates.

To ward off the summer heat, Renshade is a low-cost aluminium product that can be reversibly applied to windows. This can be a lifesaver in some properties where the windows are exposed to direct sun and reflected radiant heat.

SEE ALSO *Windows and window coverings, Renshade (page 158)*

Minimising electricity use and costs

In some states and territories around Australia, electricity monitoring devices are available for free or at low cost. However, to use these the property may need an accessible smart meter located close enough to the dwelling for bluetooth to connect to an app on your phone.

Some local councils and libraries loan out home energy kits. These may include what's called a power meter, which can be used to see how much electricity an individual appliance, such as an electric heater, uses.

There are many resources available, from both government bodies and not-for-profit organisations, that focus on helping renters to reduce household energy bills.[2]

SEE ALSO *Minimising your electricity use and costs (page 169) – many of the tips here can be applied to rental properties – and Further reading and watching (page 200) for specific resources*

Can renters benefit from solar PV?

Can renters benefit if solar PV panels are installed on a rental property?

Some renters live in homes with solar PV installed and directly reap the benefits in lower electricity bills. To encourage the installation of more solar PV on rentals, the Victorian government currently has a Solar for Rentals program. The Solar Victoria website features templates describing how both the landlord and the tenants can share the benefits of installing solar PV panels on a rental property.[3]

The not-for-profit organisation Solar Savers provides free advice to landlords and tenants whose rental property is in participating council areas. The organisation Better Renting is lobbying governments to require that landlords install solar PV panels on tenants' properties where possible.

SEE ALSO *Solar PV panels and batteries: are they worth it?* *(page 177)*

Afterword

On the issue of climate change, we are past the denial but sadly not yet up to the urgent task. Thankfully there are people, through passion, interest or concern, who have been practising the future for decades already. They are sometimes called early adopters but, for me, they are heroes and friends who enjoy the challenge of looking forward and prototyping their life around what life could be and how it could be better. They take adventurous journeys before they are necessarily the easy or the cheap thing to do, and they do it, quite honestly, for the challenge, the love and the thrill. Tim Forcey convened more than 116,000 such people online in the extraordinary background research – and more importantly the practice! – of the ideas in this book.

We are so lucky to have these heroes amongst us, and even luckier that they write books like this, full of hard-earned knowledge and wisdom. Though that sounds grand, it really is what all of these tips and tricks on how to build yourself an electric and efficient home are. Tim has pioneered a way and now generously gives us his guidebook, which will save us time and money and bring us to his destination – the future – at a discount.

No one's electrification journey will be the same. Yes, we drive cars, live in houses, cook meals and take hot showers, but our own desires, the different places we live and the buildings we inhabit all have their own characteristics and challenges. So perhaps even more important than Tim's personal journey is his pivotal role in building community – a community of early adopters who for a decade have been sharing local tips and tricks that aren't prescriptive but now represent a body of knowledge that is sympathetic to everyone's idiosyncrasies.

We have two decades to turn this climate problem around. That means we need to be practically at zero emissions in twenty years. Don't forget to apply the same principles this book espouses to also electrify your vehicles and your small businesses.

This great little book will be a very practical guide for so many of us.

Dr Saul Griffith, co-founder Rewiring Australia

Further reading and watching

I have written articles, appeared in videos and conducted presentations on every topic in this book. They are searchable online. Here are some other valuable resources.

Books

The Big Switch: Australia's electric future, Saul Griffith, Black Inc., 2022.

The Energy-Freedom Home: How to wipe out electricity and gas bills in nine steps, Beyond Zero Emissions, Scribe Publications Pty Ltd, 2015.

Your Home: Australia's guide to environmentally sustainable homes, Australian Government, 2021. (Available from yourhome.gov.au)

Online

Better Renting Coping Cookbook: betterrenting.org.au

The Conversation: Links to a collection of articles by Tim Forcey: theconversation.com/profiles/tim-forcey-104740/articles

Eco Evo: Draughty house videos with John Konstantakopoulos: ecoevo.com.au

ecoMaster: 'How to' videos with Maurice Beinat: youtube.com/ @ecoMasterAu

Electric Vehicle Consumer Hub: electricvehiclecouncil.com.au/ electric-vehicle-consumer-hub

Electrical appliance energy star ratings: energyrating.gov.au

Electrify Boroondara: Vic community organisation: electrifyboroondara.org

Electrify 2515: NSW community organisation with Dr Saul Griffith: electrify2515.org

Energy Tips: Interactive website and videos by Geelong Sustainability: energytips.org.au

Environment Victoria: Information for renters:
environmentvictoria.org.au/sustainability-hub/renters

FairAir: The FairAir website aims to provide in-depth, unbiased technical
information to consumers about home cooling options
and products and includes interactive guides and calculators:
fairair.com.au

The Fifth Estate: Australia's leading online newspaper for the sustainable
built environment: thefifthestate.com.au

Getting Off Gas Toolkit: gettingoffgastoolkit.com

Green It Yourself: 'How to' videos with Lish Fejer:
greenityourself.com.au

Light House Architecture and Science: online journal by Jenny Edwards:
lighthouseteam.com.au/journal

Make the Switch: website and calculator by the ACT Conservation
Council: maketheswitch.org.au

My Efficient Electric Home (MEEH) Facebook group:
facebook.com/groups/MyEfficientElectricHome

New Energy Thinking: online blog by Richard Keech:
newenergythinking.com

Renew: A not-for-profit organisation that produces *Renew Magazine*,
Sanctuary Magazine, and all the buyers guides for topics mentioned
in this book for home owners and renters, as well as other online
resources: renew.org.au

Renew Economy/Switched On: Information, news, podcasts:
switchedon.reneweconomy.com.au

Rewiring Australia: Dr Saul Griffith: rewiringaustralia.org

Scorecard: Find a scorecard assessor to rate your home's energy
efficiency: homescorecard.gov.au

Solar Analytics: Information on monitoring solar use and generation:
solaranalytics.com.au

Solar for renters: Victorian government information on solar for renters:
solar.vic.gov.au/information-renters

Solar Quotes: Information on solar PV and batteries: solarquotes.com.au

Solar savers for renters: Provides free advice to landlords and tenants:
solarsavers.org.au/rentals

Suburban Earthship: Website with many tips on electrifying your home
efficiently: suburbanearthship.wordpress.com

Sustainable House Day: sustainablehouseday.com

Windows Energy Rating Scheme (WERS): awa.associationonline.com.au

Zero Carbon Merri-bek: Leading Melbourne local council: zerocarbonmerri-bek.org.au/go_all_electric

Useful products

Allume Energy: allumeenergy.com.au

ecoGlaze secondary glazing: ecoglaze.com.au

ecoMaster: ecomaster.com.au

Magnetite secondary glazing: magnetite.com.au

Renshade: renshade.com.au

Retrofit Double Glaze: retrofitdoubleglaze.com.au

SealaSash: sealasash.com.au

Showerdome: showerdome.com.au

Stopnoise: stopnoise.com.au

Thermawood: thermawood.com.au

TwinGlaze: twinglaze.com.au

Acknowledgements

This book was written on the traditional lands of the Bunurong people. I thank and pay respect to the Elders of these communities, past and present.

With all the research reports and articles I had written on this topic, the endless hours I was spending blogging and commenting online, and the time I was spending in hundreds and hundreds of clients' homes saying a lot of the same messages over and over, I often thought, 'Should I just put this all in a book?' A book might be one last contribution I could make to help fight the climate crisis.

But being 65 years old and having been climate active for nearly twenty years, there were other things I was thinking of doing with my remaining years. I eventually decided that I didn't have the energy to sell the idea of a book to a publisher. But one day Jane Willson from Murdoch Books contacted me and asked if I wanted to write a book.

Jane may have been surprised by how quickly I said, 'Okay!'

If this is one more crazy way to spread the word about how my fellow Australians could save some money in their homes, be healthier and more comfortable, while at the same time addressing the big thing – global climate impacts – I've got to give it a try. So, Jane, thanks for asking me if I wanted to write a book. And thanks to the team at Murdoch Books for bringing it to life.

Thanks to Alan Pears who has been my guru since 2006 and who has never forgotten anything that has ever happened in the energy space in Australia, long before I landed here.

Thanks to Matthew Wright who, when leading Beyond Zero Emissions, was a source of youthful inspiration. It was at BZE that I met many other staff and volunteers including Phillip Sutton, Richard Keech, Mark Ogge, Heidi Lee, Trent Hawkins, Robin Gardner and dozens more, a ragtag group who decided in 2007 to take on a massive task that Australian governments or academic institutions should have already been doing. They're still trying to catch up with BZE.

Thanks to Roger Dargaville who, when at the University of Melbourne Energy Institute in 2015, agreed it would be a very useful thing to work out which is the cheapest way to heat.

Thanks again to Richard Keech and to the other admins at the MEEH Facebook group for keeping our 116,000 (and growing) members moving in a positive direction: Katy Daily, Alessandra Whiting, Sophia MacRae, Talina Edwards and Simon Samson.

Thanks to Jenny Edwards, owner of Light House Architecture and Science in Canberra, who was able to find time to be a MEEH admin, offer advice freely to our members and teach an industrial chemical engineer so much about building science.

Thanks to thermal envelope improvement experts Lyn and Maurice Beinat, owners of the business ecoMaster, for working tirelessly in this space for over twenty years, and for putting all of your intellectual property up on the internet for the world to use. It looks like we may finally have found some traction.

As I wrote in the Dedication, thanks especially to my wife of 40 years. Thanks to my kids and their partners, who have supported me through every crazy thing.

And a final thanks also to the thousands of people I've met on this journey, and to the people I continue to meet every day, working hard, volunteering hard and donating hard to make this world a better place.

Endnotes

Introduction: why live in an efficient electric home?

1. *'Further, I summarise the countless discussions I've had with householders online ...'* <facebook.com/groups/MyEfficientElectricHome>

2. *'At MEEH, we've influenced many millions of dollars' worth of home-improvement decisions.'* Forcey, Tim, 'Can Facebook help make your home more sustainable?' The Conversation, 10 January 2017 <theconversation.com/can-facebook-help-you-make-your-home-more-sustainable-70588>

1. The climate emergency calls for change

1. *'At BZE we were two years ahead of AEMO when we published ...'* Beyond Zero Emissions, August 2011 <bze.org.au/research/report/stationary-energy-plan>

2. *'In 2013, the team at BZE (where I continued as a volunteer) went on to publish another ...'* Beyond Zero Emissions, August 2013 <bze.org.au/research/report/energy-efficient-buildings-plan>; Beyond Zero Emissions, *The Energy-Freedom Home*, Scribe Publications, 2013.

3. *'In 2015, we were the first to show how Australian households could ...'* Forcey, Tim, 'The cheapest way to heat your home with renewable energy – just flick a switch', The Conversation, September 7 2015 <theconversation.com/the-cheapest-way-to-heat-your-home-with-renewable-energy-just-flick-a-switch-47087>

4. *'When we released our research, we received some media attention.'* Arup, Tom, 'Heat pump tech could save Victorian homes up to $658 a year on gas: a report', *The Age*, 26 August 2015 <theage.com.au/national/victoria/heat-pump-tech-could-save-victorian-homes-up-to-658-a-year-on-gas-report-20150825-gj7gzt.html>

5. *'At the AEF I was involved with the development of the Victorian government's computer-based ...'* <energy.vic.gov.au/households/save-energy-and-money/residential-efficiency-scorecard>

6. *'Medical data highlights that many premature deaths ...'* Barlow, Cynthia Faye; Barker, Emma; Daniel, Lyrian, 'This unhealthy housing

problem could be affecting you' SBS News, 17 May 2017 <sbs.com.au/
news/article/this-unhealthy-housing-problem-could-be-affecting-you/
560f5001l>

2. Electrifying our homes is now a thing

1. *'The homes of BZE colleagues Matthew Wright, Richard Keech and
John Shiel ...'* Beyond Zero Emissions, *The Energy-Freedom Home*,
Scribe Publications, 2015.

2. *'Entire new housing estates are being built with no connection to
any gas grid ...'* Burgess, Katie, 'Ginninderry to be first Canberra
suburb without natural gas', *The Canberra Times*, 24 April 2018,
<canberratimes.com.au/story/6023308/ginninderry-to-be-
first-canberra-suburb-without-natural-gas/; Johnstone, Henry,
'Sustainable coastal estate helping residents to slash energy
bills', realestate.com.au, 7 August 2023 <realestate.com.au/news/
sustainable-coastal-estate-helping-residents-to-slash-energy-bills>

3. *'Dozens of local councils, led by the more progressive ones ...'* Zero Carbon
Merri-bek website <zerocarbonmerri-bek.org.au/energy-switch/
go-all-electric>

4. *'The government of the Australian Capital Territory (ACT) is enacting
policies ...'* Vorrath, Sophie, 'ACT passes first law in Australia banning
gas in new homes as fossil empire strikes back', Renew Economy,
8 June 2023 <reneweconomy.com.au/act-passes-first-law-in-australia-
banning-gas-in-new-homes-as-fossil-empire-strikes-back>

5. *'The Victorian government has declared that as of 1 January 2024 ...'*
Dorrington, Benn, 'New home electrification puts the heat on
cooking with gas', realestate.com.au, <realestate.com.au/news/
new-home-electrification-puts-the-heat-on-cooking-with-gas>

6. *'Independent federal parliamentarians are pushing for all Australian
states ...'* Rae, Marion, 'Electrifying call for national ban on gas to new
homes', *The Canberra Times*, 21 June 2023 <canberratimes.com.au/
story/8242249/electrifying-call-for-national-ban-on-gas-to-new-homes>

7. *'The Australian government announced in its May 2023 budget ...'* Renew,
'Australia's first electrification budget – What the 2023 Federal budget
means for households', Renew.org.au, 11 May 2023 <renew.org.au/our-
news/australias-first-electrification-budget-what-the-2023-federal-
budget-means-for-households>

8. *'These distributors have submitted funding requests to the regulator ...'*
Vorrath, Sophie, 'Gas networks promise "green" home supply, but
put their money on electrification', 18 August 2023, Renew Economy,

<reneweconomy.com.au/gas-networks-promise-green-home-supply-but-put-their-money-on-electrification>

9. '*An afternoon presenter on Melbourne 3AW radio said ...*' Author's record, 19 September 2023.

10. '*... a poll I ran identified eleven barriers to people electrifying their homes ...*' Forcey, Tim, 'Poll on electric homes – 11 barriers that home owners say is stopping them', The Fifth Estate, 9 November 2023 <thefifthestate.com.au/columns/spinifex/poll-on-electric-homes-11-barriers-that-home owners-say-is-stopping-them>

11. '*Then we started a long and drawn-out comfort and energy makeover ...*' Forcey, Tim, 'Home electrification in one chart', The Fifth Estate, 25 August 2022 <thefifthestate.com.au/columns/spinifex/home-electrification-in-one-chart>

12. '*We obtained those panels as part of a community bulk-buy ...*' 'EKO Energy', Sustainable Australia Fund <sustainableaustraliafund.com.au/eko-energy>

13. '*The demand for grid-supplied electricity has actually been falling ...*' Forcey, Tim, 'Electrifying our homes – Can the gas industry stop us? The Fifth Estate, 16 August 2021 <thefifthestate.com.au/energy-lead/energy/electrifying-our-homes-can-the-gas-industry-stop-us>

3. Stop burning stuff!

1. '*Instead, they turned to increasingly complex ...*' Forcey, Tim, Renew Economy, 20 August 2021 <reneweconomy.com.au/households-want-gas-fossil-fuel-industry-telling-tales-as-it-fights-for-its-future>

2. '*The harmful consequences are such that fracking has often been featured ...*' May, Natasha, 'Fracking projects in NT risk exposing people to cancer and birth defects, report finds', The Guardian, 4 September 2023 <theguardian.com/australia-news/2023/sep/04/fracking-projects-in-nt-risk-exposing-people-to-cancer-and-birth-defects-report-finds>

3. '*Starting in 2015, the new access to international buyers ...*' Forcey, Tim, 'Heading north: how the export boom is shaking up Australia's gas market', The Conversation, 18 January 2016 <theconversation.com/heading-north-how-the-export-boom-is-shking-up-australias-gas-market-52963>

4. '*The legacy eastern state gas fields ...*' Forcey, Tim, 'Goodbye to Bass Strait Gas', Renew Magazine, 15 April 2020 <renew.org.au/renew-magazine/renewable-grid/goodbye-to-bass-strait-gas>

5. '*Gas importers are even developing plans to import expensive liquefied ...*' 'Andrew Forrest's Port Kembla LNG import vision put to the test',

The Australian Financial Review, 14 September 2023 <afr.com/
chanticleer/andrew-forrest-s-port-kembla-lng-import-vision-put-
to-the-test-20230914-p5e4md>

6. *'This is an idea that just a few years ago was viewed as being absolutely
   crazy.'* Audience reaction at the March 2016 Australian Domestic Gas
   Outlook conference in Sydney, as recorded by the author.

7. *'... leaks into our Earth's atmosphere during the stages of production,
   transport by pipeline and end use.'* Forcey, Tim, 'Greehouse gas
   footprint: What we do and don't know about gas', *Renew Magazine,*
   18 July 2018 <renew.org.au/renew-magazine/climate-change/
   greenhouse-gas-footprint-of-gas>

8. *'Australia is no exception, with studies finding that the methane emissions
   from fossil gas production ...'* Foley, Mike, 'Australia vastly underreporting
   methane pollution, report finds', *The Sydney Morning Herald,* 5 July 2023
   <smh.com.au/politics/federal/australia-vastly-underreporting-methane-
   pollution-report-finds-20230704-p5dll7.html>

9. *'Personally, I have travelled to the Queensland coal seam gas fields ...'*
   <youtube.com/watch?v=Jx-jWcRzndw>

10. *'In Australia, gas producers and distributors are realising ...'* Vorrath,
    Sophie, 'Gas networks promise "green" home supply, but put
    their money on electrification', Renew Economy, 18 August 2023
    <reneweconomy.com.au/gas-networks-promise-green-home-supply-
    but-put-their-money-on-electrification>

11. *'However, it's now been shown around the world that this was simply
    a tactic ...'* Forcey, Tim, 'Hydrogen in the gas grid is a dumb
    idea', Renew Economy, 24 August 2022 <reneweconomy.com.au/
    hydrogen-in-the-gas-grid-is-a-dumb-idea-very-dumb>

12. *'Medical studies have shown there is no level at which it is safe to breathe
    wood smoke.'* Asthma Australia, 'Don't underestimate the health
    dangers of woodsmoke', 9 July 2018 <asthma.org.au/about-us/media/
    dont-underestimate-the-health-dangers-of-wood-smoke>

13. *'Recently, the fossil LPG supply industry has started talking about a greener
    LPG option ...'* *Tasmanian Times,* 13 March 2023 <tasmaniantimes.com/
    2023/03/tasmanian-lpg-to-begin-renewable-transition-in-2025>

14. *'It's not clear when, if at all, bio LPG will appear in Australia ...'*
    Law, Kelvin, 'What's the problem with biofuels?' Channel New
    Asia, 2 January 2024 <channelnewsasia.com/commentary/
    cop28-fossil-fuel-biofuel-sustainable-aviation-palm-oil-4016206>

4. Where our electricity comes from, now and in the future

1. *'In 2023, 39 per cent of Australian electricity came from renewables ...'* Data sourced from opennem.org.au.

2. *'This was up from 35 per cent the year before.'* Data sourced from opennem.org.au.

3. *'In fact, the current Australian government has indicated ...'* Department of Climate Change, Energy , the Environment and Water <dcceew.gov. au/energy/strategies-and-frameworks/powering-australia>

4. *'The Australian Energy Market Operator (AEMO) modelled one scenario where electricity generation ...'* Edis, Tristan; Bowyer, Johanna, 'Afraid of the dark and clutching at coal, ESB is still beating capacity market drum', Renew Economy, 8 October 2021 <reneweconomy.com.au/afraid-of-the-dark-and-clutching-at-coal-esb-is-still-beating-capacity-market-drum>

5. *'It is a remarkable fact that in Australia today we require less electricity ...'* Data sourced from opennem.org.au.

6. *'One reason for this is because newer appliances use less electricity than older versions.'* Sandiford, Mike; Forcey, Tim; Pears, Alan; McConnell, Dylan, 'Five years of declining annual consumption of grid-supplied electricity in Eastern Australia: causes and consequences', Elsevier, 2015 < climatecollege.unimelb.edu.au/files/site1/docs/949/Five%20 Years%20of%20Declining%20Annual%20Consumption%20of%20Grid-Supplied%20Electricity%20in%20Eastern%20Australia.pdf>

5. Heating and cooling

1. *'But don't forget about ways you can improve ...'* Your Home, 'Passive design', Australian Government, Australia's Guide to Environmentally Sustainable Homes, Your Home <yourhome.gov.au/passive-design>

2. *'I came to realise this was a very big cost-of-living win ...'* Forcey, Tim, 'Explainer: What's the cheapest way to heat my house if I get off gas?' Renew Economy, 28 May 2021 <reneweconomy.com.au/ explainer-whats-the-cheapest-way-to-heat-my-house-if-i-get-off-gas>

3. *'As reported in 2015 in* The Age, *our research found ...'* Arup, Tom, 'Heat pump tech could save Victorian homes up to $658 a year on gas: a report', *The Age*, 26 August 2015 <theage.com.au/national/victoria/ heat-pump-tech-could-save-victorian-homes-up-to-658-a-year-on-gas-report-20150825-gj7gzt.html>

4. *'Heat pumps aren't magic, but they can be mysterious.'* Forcey, Tim, 'Heat pumps, part 2, space heating', June 2020.

5. *'To ensure poisonous carbon monoxide gas isn't entering your home, gas heaters must be checked ...'* Energy Safe Victoria, 19 September 2023 <esv.vic.gov.au/community-safety/energy-safety-guides/home-safety/heating-your-home-gas>

6. *'Secondly, ducted heating and cooling systems can be less efficient and more costly to operate than ductless systems.'* Keech, Richard, 'Don't do ducts', New Energy Thinking, 20 October 2018 <newenergythinking.com/2018/10/20/dont-use-ducts>

7. *'... a centrally ducted system is meant to be operated as a closed-loop system.'* Forcey, Tim, 'Seven Things People Don't Know About Their Homes', The Fifth Estate, 28 June 2022 <thefifthestate.com.au/housing-2/7-things-people-dont-know-about-their-homes>

8. *'This can be useful not only for aesthetic reasons ...'* Keech, Richard, 'Multi-head splits versus single split-systems', New Energy Thinking, 18 May 2018, <newenergythinking.com/2018/06/18/to-split-or-not-to-split>

9. *'... all that should be needed in winter or summer is a bit of reverse-cycle air conditioning.'* <facebook.com/groups/MyEfficientElectricHome/posts/1409920739052365>

10. *'Recent research at RMIT has confirmed why.'* Pears, Alan; Willand, Nicola; Vahaji, Sara; Moore, Trivess, 'Replacing gas heating with reverse-cycle aircon leaves some people feeling cold. Why? And what's the solution?' The Conversation, 2 October 2023, <theconversation.com/replacing-gas-heating-with-reverse-cycle-aircon-leaves-some-people-feeling-cold-why-and-whats-the-solution-213542>

11. *'Additionally, many households these days find that the humidified feel ...'* Forcey, Tim, 'Evaporative cooling for your home – Have the benefits drifted away?' 21 September 2021 <thefifthestate.com.au/columns/spinifex/evaporative-cooling-for-your-home-have-the-benefits-drifted-away>

12. *'Thankfully, in Victoria this will change as new laws restrict the use of gas in new-build homes.'* Press release from the office of Hon. Jacinta Allan MP <premier.vic.gov.au/new-victorian-homes-go-all-electric-2024>

13. *'They can cool a small space where you are sitting...'* 'Portable air conditioners: Why you shouldn't like them', <youtube.com/watch?v=_-mBeYC2KGc>

6. Heat your water the cheapest way

1. *'... consider heating your water with a hot-water heat pump.'* Forcey, Tim, 'Heat pumps, part 1, hot water', November 2019 <youtube.com/watch?v=XGmxqwFBirI>

2. *'I liken buying a hot-water heat pump to buying a car.'* Forcey, Tim, 'So you want to by a heat pump? Here's how', Renew Economy, 19 November 2023 <switchedon.reneweconomy.com.au/content/so-you-want-to-buy-a-heat-pump-heres-how>

3. *'The consumer product reviewers CHOICE have on their website free technical specifications ...'* Choice, 'Heat pump hot water systems comparison', 16 May 2023 <choice.com.au/home-improvement/water/hot-water-systems/review-and-compare/heat-pump-hot-water-systems>

4. *'... there are other electrical water heating options with small footprints.'* Turner, Lance, 'Hot water buyer's guide' *Renew Magazine*, 20 March 2020 <renew.org.au/renew-magazine/buyers-guides/hot-water-buyers-guide-2>

5. *'Another is when something associated with the solar-thermal part fails, such as a circulating pump ...'* Keech, Richard, 'The sad reality of commodity solar hot water', New Energy Thinking, 20 June 2021 <newenergythinking.com/2021/06/20/the-sad-reality-of-commodity-solar-hot-water>

7. Bye-bye gas grid: cooking without gas

1. *'When you are finished cooking and you shut off the cooktop ...'* Forcey, Tim, 'How electric induction cookers work and why they clear the air', Renew Economy, 30 May 2021, <reneweconomy.com.au/explainer-how-electric-induction-cookers-work-and-why-they-clear-the-air>

2. *'... those other chemicals and fine particulates listed above can have a range of bad health impacts including asthma ...'* Brambrick, Hilary; Charlesworth, Kate; Bradsaw, Simon, 'Kicking the gas habit: How gas is harming our health,' Climate Council, 6 May 2021 <climatecouncil.org.au/resources/gas-habit-how-gas-harming-health>

3. *'The nitrous oxides produced when cooking with gas ...'* Asthma Australia, 'Asthma triggers' <asthma.org.au/triggers/gas-appliances>

4. *'The Victorian government plans to offer rebates for households installing induction cooktops.'* Victorian State Government, 'Gas substitution Roadmap Update' <energy.vic.gov.au/__data/assets/pdf_file/0027/691119/Victorias-Gas-Substitution-Roadmap-Update.pdf>

5. *'For more on this topic, you can read an excellent article ...'* Edis, Tristan, 'If I go all electric do I need to upgrade my grid connection?' 20 April 2023, *The Australian Financial Review* <afr.com/companies/energy/if-i-go-all-electric-do-i-need-to-upgrade-my-grid-connection-20230419-p5d1pn>

6. *'The British Heart Foundation recommends a distance of 60 centimetres.'* The British Heart Foundation, 'Can I use an induction hob if I have a pacemaker?' <bhf.org.uk/informationsupport/heart-matters-magazine/medical/ask-the-experts/induction-hobs-and-pacemakers>

7. *'They might also need to be wary of hair dryers ...'* Bowring, Declan, ABC News website, 19 January 2023 <abc.net.au/news/2023-01-19/pacemaker-users-stay-away-from-induction-stoves/101872736>

8. *'The Victorian regulator announced in 2023 that individual households ...'* Forcey, Tim; Pears, Alan, 'Great Barrier Fee: energy regulator slugs electrified homes', The Fifth Estate, 13 June 2023 <thefifthestate.com.au/columns/spinifex/great-barrier-fee-energy-regulator-slugs-electrified-homes>

8. Do you live in a leaky bucket?

1. *'The 7-star requirement for new-build homes supersedes the 6-star requirement introduced ...'* National Construction Code, 'Building for 7 stars: top tips and guidance', 26 October 2022 <ncc.abcb.gov.au/news/2022/building-7-stars-top-tips-and-guidance>

2. *'Many older homes rate 2 stars or less ...'* Barnett, Adam, 'Cold weather is a bigger killer than extreme heat – here's why', The Conversation, 22 May 2015 <theconversation.com/cold-weather-is-a-bigger-killer-than-extreme-heat-heres-why-42252>

3. *'The ACT's Energy Efficiency Rating (EER) scheme is one example of mandatory disclosure ...'* ACT Government/Planning 'Energy efficiency' <planning.act.gov.au/professionals/regulation-and-responsibilities/responsibilities/energy-efficiency>

4. *'The United Kingdom and some other places around the world also have mandatory disclosure.'* 'Buying or selling your home: energy performance certificates', Government UK <gov.uk/buy-sell-your-home/energy-performance-certificates>

5. *'In my view, thermal mass is overrated.'* Keech, Richard, 'Is thermal mass overrated?', New Energy Thinking, 16 June 2018 <newenergythinking.com/2018/06/16/is-thermal-mass-over-rated>

9. Draught-proofing and clearing the air

1. *'A draughty home can let air in that's polluted with smoke from a distant bushfire ...'* Twyford, Lottie, 'Australians struggled to breathe during the Black Summer bushfires. If the smoke comes back, are we any better prepared?', ABC News, 12 January 2024 <abc.net.au/news/2024-01-12/canberra-choked-through-black-summer-bushfire-

smoke-preparation/103154116>; ecoMaster, 'Are smoke, dust, pollen or draughts getting inside your home?' <ecomaster.com.au/bushfire-and-indoor-air-quality>

2. *'Secondly, a leaky home can allow moisture ...'* EPA, 'A brief guide to mold, moisture and your home', United States Environmental Protection Agency <epa.gov/mold/brief-guide-mold-moisture-and-your-home#tab-7>

3. *'Many of Renew's buying guides are available on the internet.'* Turner, Lance, 'A draught sealing buyers guide', *Renew Magazine*, 3 April 2019 <renew.org.au/renew-magazine/buyers-guides/draught-sealing-guide>

4. *'The founders of the draught-proofing company ecoMaster ...'* ecoMaster, Video library <ecomaster.com.au/video-library>

5. *'... in some parts of the world, even radioactive contaminants.'* Australian Radiation Protection and Nuclear Safety Agency, 'Radon map of Australia', Australian Government <arpansa.gov.au/understanding-radiation/radiation-sources/more-radiation-sources/radon-map>

6. *'Among Australian housing professionals, discussions are beginning about whether continuous ventilation ...'* Zeller, Alex, 'Only fans a draft guide to continuous ventilation in homes in Australia and New Zealand', LinkedIn, 19 October 2023 <linkedin.com/pulse/only-fans-draft-guide-continuous-ventilation-new-zealand-alex-zeller-v4qae>

7. *'The recommended relative humidity (RH) range is between ...'* Forcey, Tim, 'Moisture management, and the death of an Aussie icon', The Fifth Estate, 22 May 2023 <thefifthestate.com.au/energy-lead/moisture-management-and-the-death-of-an-aussie-icon>

8. *'... and mould, which then leads to a number of health issues including asthma.'* Sterling, E, M; Arundel, A, 'Criteria for Human Exposure to Humidity in Occupied Buildings', *ASHRAE Transactions*, Vol 91, Part 1, 1985.

9. *'This can lead to excessive moisture being added to the air.'* Forcey, Tim, 'Moisture management, and the death of an Aussie icon', The Fifth Estate, 22 May 2023, <thefifthestate.com.au/energy-lead/moisture-management-and-the-death-of-an-aussie-icon>

10. *'The health impacts of elevated carbon dioxide continue to be studied.'* Odgen, Chris, 'CO_2 affects human health at lower levels than previously thought', airqualitynews.com, 10 July 2019 <airqualitynews.com/health/co2-affects-human-health-at-lower-levels-than-previously-thought>

11. *'At 1,000 ppm and beyond you may have feelings of restlessness …'* Airthings, 24 July 2023 <airthings.com/resources/carbon-dioxide-co2-indoor-air-home>

12. *'A German practice of routinely allowing a large flush of air through the home …'* German word of the day: *Stoβlüften', The Local*, 29 December 2022 <thelocal.de/20190117/stolften>

13. *'… fine particulates that are linked to heart and lung disease …'* NSW Health, 'Particulate matter (PM10 and PM2.5)', <health.nsw.gov.au/environment/air/Pages/particulate-matter.aspx>

10. Insulation: a priority

1. *'… a minimum legal requirement to have insulation applied to each of those parts of a home.'* For more information on insulation, see Turner, Lance, 'The key to thermal performance: insulation buyers guide', *Renew Magazine*, 27 June 2017 <renew.org.au/renew-magazine/buyers-guides/insulation-buyers-guide>

2. *'… give off a bit of light and ran hot enough to cook a chicken, have caused many fires.'* NSW Fire and Rescue, 'Halogen downlights' 2023 <fire.nsw.gov.au/page.php?id=709#:~:text=Halogen%20down%2Dlights%20create%20a,get%20anywhere%20up%20to%20370C>

3. *'A whirlybird is a moving piece of equipment …'* Keech, Richard, 'Roof ventilation – not a fan', New Energy Thinking, 27 December 2020 <newenergythinking.com/2020/12/27/roof-ventilation-not-a-fan>

4. *'… but one tip is to see if the source of the moisture entering a roof space can be eliminated.'* Edwards, Jenny, 'Condensation management', *Renew Magazine*, 24 April 2019, <renew.org.au/renew-magazine/efficient-homes/condensation-management>

5. *'Some owners of these have opted to install insulation and plasterboard inside …'* Stapleton, James. 'DIY: Insulating raked ceilings', *Renew Magazine*, 31 October 2019, <renew.org.au/renew-magazine/diy/diy-insulating-raked-ceilings>

11. Windows and window coverings

1. *'Installing excellent window coverings inside your home, and then actually using them …'* Cumming, Anna, 'Not just window dressing: High peformance curtains and blinds', *Renew Magazine*, 29 March 2018 <renew.org.au/renew-magazine/buyers-guides/high-performance-curtains>

2. *'I've measured the temperature of decks in the midday sun ...'* Cumming, Anna, 'Keep your cool with external shading', *Renew Magazine*, 27 November 2017 <renew.org.au/sanctuary-magazine/ideas-advice/keep-your-cool-with-external-shading>

3. *'If you've done as much as you can with window coverings inside and ...'* 'Windows that perform', *Renew Magazine*, 26 April 2018 <renew.org.au/renew-magazine/buyers-guides/window-buyers-guide>

4. *'Windows are costly, both in terms of money and ...'* The Lighthouse Team, 'Windows 102: ratios and coefficients', 26 May 2018 <lighthouseteam.com.au/windows-101/windows-102-ratios-and-coefficients>

12. Minimising electricity use and costs

1. *'Victoria has had these since the roll-out began back in 2006.'* The Victorian Auditor-General's Office, 'Realising the benefits of smart meters', 16 September 2015 <audit.vic.gov.au/report/realising-benefits-smart-meters?section=2>

2. *'There you will find access to a file of half-hourly data ...'* Jemena Electricity Outlook <electricityoutlook.jemena.com.au>

3. *'A few government programs and some electricity suppliers have offered these for free or at low cost.'* Victorian State Government, 'In-home displays', Energy, Environment and Climate Action, updated 22 September 2023 <energy.vic.gov.au/households/victorian-energy-upgrades-for-households/in-home-displays>

4. *'This will cost extra for the equipment involved ...'* Solar Quotes, 'A monitoring system for your solar' <solarquotes.com.au/good-solar-guide/monitoring-systems>

5. *'There are comparison websites set up by state and federal governments...'* <energymadeeasy.gov.au>

6. *'Wholesale electricity prices in eastern Australia can range ...'* The Australian Energy Market Commission media release, 7 December 2023 <aemc.gov.au/news-centre/media-releases/new-rule-ensure-reliability-energy-system>

7. *'We'll see more and more of this sort of demand–response activity going on.'* Williamson, Rachel, 'How to unlock $250 billion in energy assets under (and on) our own roofs', Renew Economy, 13 October 2023 <reneweconomy.com.au/how-to-unlock-250-billion-in-energy-assets-under-and-on-our-own-roofs>

8. *'Many government bodies and sustainability oriented not-for-profit organisations ...'* Sustainability Victoria, 'Save energy in the home', updated 2 August 2023 <sustainability.vic.gov.au/energy-efficiency-and-reducing-emissions/save-energy-in-the-home>

13. Solar PV panels and batteries: are they worth it?

1. *'With around one-third of Australian homes now having solar PV panels ...'* Morton, Adam, 'Go hard and go big': How Australia got solar panels onto one in every three houses', *The Guardian Australia*, 2 November 2023 <theguardian.com/environment/2023/nov/01/how-generous-subsidies-helped-australia-to-become-a-leader-in-solar-power>

2. *'In 2016, I wrote an article with the cheeky title ...'* Forcey, Tim, '22 ways to cut your energy bills (before spending on solar panels)', The Conversation, 15 June 2016 <theconversation.com/22-ways-to-cut-your-energy-bills-before-spending-on-solar-panels-60847>

3. *'... sell your electricity only at a price as low as $0.03 per kilowatt-hour.'* Vorrath, Sophie, 'Rooftop solar feed-in tariffs to be slashed another 33 pct as duck curve grows', One Step Off the Grid, 22 November 2023 <onestepoffthegrid.com.au/rooftop-solar-feed-in-tariffs-to-be-slashed-another-33-pct-as-duck-curve-grows>

4. *'Unfortunately, electricity distributors setting a constant ...'* Williamson, Rachel, 'Flexible export limits: The next phase for rooftop solar kicks off in Australia', Renew Economy, 6 July 2023 <reneweconomy.com.au/flexible-export-limits-the-next-phase-for-rooftop-solar-kicks-off-in-an-australia-first>

5. *'When you are considering investing in solar PV ...'* Sykes, Jeff, 'Solar system size limits: How much does your local network allow?' Solar Choice, 8 August 2023, <solarchoice.net.au/learn/design-guide/solar-system-size-limits-by-network>

6. *'... the electrical energy stored in the millions of our vehicle batteries ...'* Gaton, Bryce, 'EV plug wars and a reality check for vehicle to grid technology', The Driven, 20 January 2023 <thedriven.io/2023/01/20/ev-plug-wars-and-a-reality-check-for-vehicle-to-grid-technology>

7. *'... a SolShare project in Highett, Victoria, distributes electricity from a single solar-PV system ...'* Vorrath, Sophie, 'Melbourne apartment complex switches on shared solar system', One Step Off the Grid, 25 May 2018 <onestepoffthegrid.com.au/melbourne-apartment-complex-switches-shared-solar-system>

14. I'm a renter, what can I do?

1. *'Renters can also view an article ...'* Norman, James, 'Bubble wrap insulation and DIY air-con: how renters can keep their home cool during summer', *The Guardian Australia*, 6 January 2024 <theguardian.com/australia-news/2024/jan/06/australian-summer-keep-home-cool-without-air-conditioning-#>

2. *'Many government bodies and sustainability oriented not-for-profit organisations ...'* Environment Victoria <environmentvictoria.org.au/sustainability-hub/renters>

3. *'The Solar Victoria website features templates ...'* Victorian Government, Solar Victoria <solar.vic.gov.au/information-renters>

Index